THE CONSTANTS
OF NATURE

John D. Barrow is Research Professor of Mathematical
Sciences in the Department of Applied Mathematics and
Theoretical Physics at the University of Cambridge. He
has recently been appointed the Gresham Professorship
in Astronomy, one of the oldest and most esteemed
professorships in science in the UK. He is the author
of several bestselling books, including *Theories of
Everything, Impossibility* and *The Book of Nothing*.

ALSO BY JOHN D. BARROW

Theories of Everything

The Left Hand of Creation
(With Joseph Silk)

L'Homme et le Cosmos
(With Frank J. Tipler)

The Anthropic Cosmological Principle
(with Frank J. Tipler)

The World Within the World

The Artful Universe

Pi in the Sky

Perché il mondo è matematico?

Impossibility

The Origin of the Universe

Between Inner Space and Outer Space

The Universe that Discovered Itself

The Book of Nothing

John D. Barrow

THE CONSTANTS
OF NATURE

From Alpha to Omega

VINTAGE

Published by Vintage 2003

2 4 6 8 10 9 7 5 3 1

First published in Great Britain in 2002 by
Jonathan Cape

Vintage
Random House, 20 Vauxhall Bridge Road,
London SW1V 2SA

Random House Australia (Pty) Limited
20 Alfred Street, Milsons Point, Sydney
New South Wales 2061, Australia

Random House New Zealand Limited
18 Poland Road, Glenfield,
Auckland 10, New Zealand

Random House (Pty) Limited
Endulini, 5A Jubilee Road, Parktown 2193,
South Africa

The Random House Group Limited Reg. No. 954009
www.randomhouse.co.uk

A CIP catalogue record for this book
is available from the British Library

ISBN 0 09 928647 5

Printed and bound in Great Britain by
Bookmarque Ltd, Croydon, Surrey

To Carol

'Not the power to remember, but its very opposite, the power to forget is a necessary condition for our existence.'

Sholem Ash

Contents

Preface

Some things never change. And this is a book about those things. Long ago, the happenings that made it into histories were the irregularities of experience: the unexpected, the catastrophic, and the ominous. Gradually, scientists came to appreciate the mystery of the regularity and predictability of the world. Despite the concatenation of chaotically unpredictable movements of atoms and molecules, our experience is of a world that possesses a deep-laid consistency and continuity. Our search for the source of that consistency looked first to the 'laws' of Nature that govern how things change. But gradually we have identified a collection of mysterious numbers which lie at the root of the consistency of experience. These are the constants of Nature. They give the Universe its distinctive character and distinguish it from others we might imagine. They capture at once our greatest knowledge and our greatest ignorance about the Universe. For, while we measure them to ever greater precision, fashion our fundamental standards of mass and time around their invariance, we cannot explain their values. We have never explained the numerical value of any of the constants of Nature. We have discovered new ones, linked old ones, and understood their crucial role in making things the way they are, but the reason for their values remain a deeply hidden secret. To search it out we will need to unpick the most fundamental theory of the laws of Nature, to discover if the constants that define them are fixed and framed by some over-arching logical consistency or whether chance still has a role to play.

Our first glimpses reveal a very peculiar situation. While some constants seem as if they will be fixed, others have the scope to be other than they are, and some seem completely untouched by everything else about the Universe. Do their values fall out at random? Could they really be different? How different could they be if life is to be possible in the Universe?

Back in 1986, my first book, *The Anthropic Cosmological Principle*, explored all the then-known ways in which life in the universe was sensitive to the values of the constants of Nature. Universes with slightly altered constants would be still-born, devoid of the potential to evolve and sustain the sort of organised complexity that we call life. Since that time, cosmologists have found more and more ways in which the Universe could exhibit variations in its defining constants; more and more ways in which life could have failed to emerge in the Universe. They have also begun to take seriously the possibility and actuality of other universes in which the constants of Nature do take different values. Inevitably, we find ourselves in a world where things fell out right. But what was the chance of that happening? Here we shall look at many of these possibilities, connecting them to the curious history of our attempts to understand the values of our constants of Nature.

Recently, one big story about the constants of Nature has produced a focus for media attention and detailed scientific research. It raises the most basic question of all: are the constants of Nature really constant after all? A new method of scrutinizing the constants of Nature over the last 11 billion years of the Universe's history has been devised by a group of us. By looking at the atomic patterns barcoded into the light that reaches us from distant quasars we can look and see what atoms were like when the light began its journey billions of years ago. So, were the constants of Nature always the same? The answer, unexpected and shocking, raises new possibilities for the Universe and the laws that govern it. This book will tell you about them.

I would like to thank Bernard Carr, Rob Crittenden, Paul Davies, Michael Drinkwater, Chris Churchill, Freeman Dyson, Vladimir Dzuba, Victor Flambaum, Yasunori Fujii, Gary Gibbons, J. Richard Gott, Jörg Hensgen, Janna Levin, João Magueijo, Carlos Martins, David Mota, Michael Murphy, Jason Prochaska, Martin Rees, Håvard Sandvik,

Wallace Sargent, Ilya Shlyakhter, Will Sulkin, Max Tegmark, Virginia Trimble, Neil Turok, John Webb, and Art Wolfe for discussions and contributions of ideas, results, and images.

I would also like to thank Elizabeth, for surviving at one stage the thought that the book might need to be retitled *A River Runs Through It*, and our three children David, Roger and Louise who were always worried that pocket-money might be a constant of Nature.

J.D.B
Cambridge, April 2002

chapter one

Before the Beginning

'What happens first is not necessarily the beginning.'

Henning Mankell[1]

SAMELINESS

'There is nothing that God hath established in a constant cause of nature, and which therefore is done everyday, but would seem a miracle, and exercise our admiration, if it were done but once.'

John Donne[2]

Change is a challenge. We live in the fastest moving period of human history. The world around us is driven by forces that make our lives increasingly sensitive to small changes and sudden responses. The elaboration of the Internet and the tentacles of the Worldwide Web have put us in instantaneous contact with computers and their owners all round the world. The threats from unchecked industrial progress have brought about ecological damage and environmental change that appears to be happening faster than even the gloomiest prophets of doom had predicted. Children seem to grow up faster. Political systems realign in new and unexpected ways more quickly and more often than ever before. Even human beings and the information they embody are facing editorial intervention by more ambitious spare-part surgery or the re-programming of parts of our genetic code. Most forms of progress are accelerating and more and more parts of our experience have become entwined in the surge to explore all that is possible.

In the world of scientific exploration the recognition of the impact of change is not so new. By the end of the nineteenth century it had been appreciated that once upon a time the Earth and our solar system had not existed; that the human species must have changed in appearance and average mental capability over huge spans of time; and that in some broad and general way the Universe should be winding down, becoming a less hospitable and ordered place. During the twentieth century we have fleshed out this skeletal picture of a changing Universe. The climate and topography of our planet is continually changing and so are the species that live upon it. Most dramatically of all, we have discovered that the entire universe of stars and galaxies is in a state of dynamic change, with great clusters of galaxies flying away from one another into a future that will be very different from the present. We have begun to appreciate that we are living on borrowed time. Cataclysmic astronomical events are common; worlds collide. Planet Earth has been hit in the past by comets and asteroids. One day its luck will run out, the shield provided so fortuitously by the vast planet Jupiter, guarding the outer reaches of our solar system, will not be able to save us. Eventually, even our Sun will die. Our Milky Way galaxy will be drawn into a vast black hole deep in its centre. Life like our own will end. Survivors will need to have changed their form, their homes and their nature to such an extent that we would be challenged to call their continued existence 'living' by our own standards today.

We have recognised the simple secrets of chaos and unpredictability which beset so many parts of the world around us. We understand our changing weather but we cannot predict it. We have appreciated the similarities between complexities like this and those that emerge from systems of human interaction – societies, economies, choices, ecosystems – and from within the human mind itself.

All these perplexing complexities rush along and seek to convince us that the world is like a runaway roller-coaster, rocking and rolling;

that everything we once held to be true might one day be overthrown. Some even see such a prospect as a reason to be suspicious of science[3] as a corrosive effect upon the foundations of human nature and certainty, as though the construction of the physical Universe and the vast schema of its laws should have been set up with our psychological fragility in mind.

But there is a sense in which all this change and unpredictability is an illusion. It is not the whole story about the nature of the Universe. There is both a conservative and a progressive side to the deep structure of reality. Despite the incessant change and dynamic of the visible world, there are aspects of the fabric of the Universe which are mysterious in their unshakeable *constancy*. It is these mysterious unchanging things that make our Universe what it is and distinguish it from other worlds that we might imagine. There is a golden thread that weaves a continuity through Nature. It leads us to expect that certain things elsewhere in space will be the same as they are here on Earth; that they were and will be the same at other times as they are today; that for some things neither history nor geography matter. Indeed, perhaps without such a substratum of unchanging realities there could be no surface currents of change or any complexities of mind and matter at all.

These bedrock ingredients of our Universe are what this book is about. Their existence is one of the last mysteries of science that has challenged a succession of great physicists to come up with an explanation for why they are as they are. Our quest is to discover what they are but we have long known only what to call them. They are the *constants of Nature*. They lie at the root of sameliness in the Universe: why every electron seems to be the same as every other electron.

The constants of Nature encode the deepest secrets of the Universe. They express at once our greatest knowledge and our greatest ignorance about the cosmos. Their existence has taught us the profound truth that Nature abounds with unseen regularities. Yet, while we have become skilled at measuring the values of these constant quantities, our inability to explain or predict their values shows how much we have still to learn about the inner workings of the Universe.

What is the ultimate status of the constants of Nature? Are they truly constant? Are they everywhere the same? Are they all linked? Could life have evolved and persisted if they were even slightly different? These are some of the issues that this book will grapple with. It will look back to the discoveries of the first constants of Nature and the impact they had on scientists and theologians looking for Mind, purpose and design in Nature. It will show what frontier science now believes constants of Nature to be and whether a future Theory of Everything, if it exists, will one day reveal the true secret of the constants of Nature. And most important of all, it will ask whether they are truly constant.

chapter two

Journey Towards Ultimate Reality

'*Franklin:* Have you ever thought, Headmaster, that your standards might perhaps be a little out of date?
Headmaster: Of course they're out of date. Standards always are out of date. That is what makes them standards.'

Alan Bennett[1]

MISSION TO MARS

'The Mars Climate Orbiter Mishap Investigation Board has determined that the root cause for the loss of the Mars Climate Orbiter spacecraft was the failure to use metric units.'

NASA Mars Climate Orbiter Mishap Investigation Report[2]

In the last week of September 1998 NASA was getting ready to hit the press agencies with a big story. The Mars Climate Explorer, designed to skim through the upper atmosphere of Mars, was about to send back important data about the Martian atmosphere and climate. Instead, it just crashed into the Martian surface. In NASA's words,

'The MCO spacecraft, designed to study the weather and climate of Mars, was launched by a Delta rocket on

December 11th, 1998, from Cape Canaveral Air Station, Florida. After a cruise to Mars of approximately 9½ months, the spacecraft fired its main engine to go into orbit around Mars at around 2 a.m. PDT on September 23, 1999. Five minutes into the planned 16-minute burn, the spacecraft passed behind the planet as seen from Earth. Signal re-acquisition, nominally expected at approximately 2:26 a.m. PDT did not occur. Efforts to find and communicate with MCO continued up until 3 p.m. PDT on September 24, 1999, when they were abandoned.'[3]

The spacecraft was 60 miles (96.6 km) closer to the Martian surface than the mission controllers thought, and $125 million disappeared into the red Martian dust. The loss was bad enough but when the cause was discovered it looked like a case for the force-feeding of humble pie. Lockheed-Martin, the company controlling the day-to-day operation of the spacecraft, was sending out data about the thrusters in Imperial units, miles, feet and pounds-force, to mission control, while NASA's navigation team was assuming like the rest of the international scientific world that they were receiving their instructions in metric units. The difference between miles and kilometres was enough to send the craft 60 miles off course on a suicidal orbit into the Martian surface.[4]

The lesson of this débâcle is clear. Units matter. Our predecessors have bequeathed us countless everyday units of measurement that we tend to use in different situations for the sake of convenience. We buy eggs in dozens, bid at auctions in guineas, measure horse races in furlongs, ocean depths in fathoms, apples in bushels, coal in hundredweight, lifetimes in years and weigh gemstones in carats. Accounts of all the standards of measurement in past and present existence run to hundreds of pages. All this was entirely satisfactory while commerce was local and simple. But as communities started to trade internationally in ancient times they started to encounter other ways of counting. Quantity was measured differently from country to country and conversion factors were needed, just as we change currency when travelling internationally today. Once international collaboration began on technical projects the stakes were raised.[5] Precision engineering requires accurate inter-comparison of standards. It is all very well telling your collaborators on the other side of the world that they need to make an aircraft component that is precisely one metre long, but how do you know that their metre is the same as your metre?

MEASURE FOR MEASURE – PAROCHIAL STANDARDS

'She does not understand the concept of Roman numerals. She thought we just fought World War Eleven.'

Joan Rivers[6]

Originally, standards of measurement were entirely parochial and anthropometric. Lengths were derived from the length of the king's arm or the span of his hand. Distances mirrored the extent of a day's journey. Time followed the astronomical variations of the Earth and Moon. Weights were convenient quantities that could be carried in the hand or on the back. Many of these measures were wisely chosen and are still with us today in spite of the official ubiquity of the decimal system. None is sacrosanct. Each is designed for convenience in particular circumstances. Many measures of distance were derived anthropomorphically from the dimensions of human anatomy. The 'foot' is the most obvious unit of this sort. Others are no longer so familiar. The 'yard' was the length of a tape drawn from the tip of a man's nose to the farthest fingertip of his arm when stretched horizontally to one side. The 'cubit' was the distance from a man's elbow joint to furthermost fingertip of his outstretched hand, and varies between about 17 and 25 of our inches (0.44–0.64 metres) in the different ancient cultures that employed it.[7] The nautical unit of length, the fathom, was the largest distance-unit defined from the human anatomy, and was defined as the maximum distance between the fingertips of a man with both hands outstretched horizontally to the side.

The movement of merchants and traders around the Mediterranean region in ancient times would have highlighted the different measures of the same anatomical distance. This would have made it difficult to maintain any single set of units. But national tradition and habit was a powerful force in resisting the adoption of another country's standards.

The most obvious problem with such units is the fact that men and women come in different sizes. Who do you measure as your standard? The king or queen is the obvious candidate. Even so, this results in a recalibration of units every time the throne changes hands. One notable response to the problem of the variation in human dimensions was that devised by David I of Scotland in 1150 to define the Scottish inch: he ordained that it was to be the *average* drawn from measurements of the width of the base of the thumbnail of three men: a 'mekill' [big] man, a man of 'messurabel' [moderate] stature, and a 'lytell' [little] man.

The modern metric system of centimetres, kilograms and litres, and the traditional 'Imperial' system of inches, pounds and pints are equally good measures of lengths, weights and volumes so long as you can measure them accurately. That is not the same thing as saying they are equally convenient, though. The metric system mirrors our counting system by having each unit ten times bigger than the next smallest. Imagine having a counting system that had uneven jumps. So, instead of hundreds, tens and units we had a counting system like that used in England for non-technical weights (like human body weights or horse-racing handicaps) with 16 ounces in one pound and 14 pounds in one stone.

The cleaning up of standards of measurement began decisively at the time of the French Revolution at the end of the eighteenth century. Introducing new weights and measures brings with it a certain upheaval in society and is rarely received with unalloyed enthusiasm by the populace. The French Revolution therefore provided an occasion to make such an innovation without adding significantly to the general upheaval of everything else.[8] The prevailing trend of political thinking at the time sided with the view that weights and measures should have an egalitarian standard that did not make them the property of any one nation, nor give any nation an advantage when it came to trading with others. The way to do this was believed to define measure against some agreed standard, from which all rulers and secondary measures would be calibrated. The French National Assembly enacted this into law on 26 March 1791, with the support of Louis XVI and the clear statement of principle submitted by Charles Maurice de Talleyrand:

> 'In view of the fact that in order to be able to introduce uniformity of weights and measures it is necessary that a natural and unchanging unit of mass be laid down, and that the only means of extending this uniformity to other nations and urging them to agree upon a system of measures is to choose a unit that is not arbitrary and does not contain anything specific to any peoples on the globe.'[9]

Two years later, the 'metre'[10] was introduced as the standard of length, defined as the ten millionth part of a quarter of the Earth's meridian.[11] Although this is a plausible way to identify a standard of length it is clearly not very practical as an everyday comparison. Consequently, in 1795, the units were directly related to specially made objects. At first the unit of mass was taken as the gram, defined to be the mass of one cubic centimetre of water at 0 degrees centigrade. Later it was superseded by the kilogram (1000 grams) defined as the mass of 1000 cubic centimetres of water at 4 degrees centigrade. Finally, in 1799, a prototype metre bar[12] was made together with a standard kilogram mass and placed in the archives of the new French Republic. Even today, the reference kilogram mass is known as the 'Kilogramme des Archives'.

Unfortunately, the new metric units were not at first successful and Napoleon reintroduced the old standards in the early years of the nineteenth century. The European political situation prevented an international harmonisation of standards.[13] It was not until New Year's Day 1840 that Louis Phillipe made metric units legally obligatory in France. Meanwhile they had already been adopted more universally in the Netherlands, Belgium and Luxembourg twenty-four years earlier, and by Greece in 1832. Britain only allowed a rather restricted use of metric units after 1864 and the USA followed suit two years later. Real progress only occurred in 1870 when the International Metre Commission was established and met in Paris on 8 August for the first time, to co-ordinate standards and oversee the making of new standard masses and lengths.[14] Copies of the standards were distributed to some of the member states chosen by the drawing of lots. The kilogram was the mass of a special cylinder, 39 mm in height and diameter, made of an alloy of platinum and iridium[15] kept under three glass bell-jars and stored inside a vault at the International Bureau of Standards in Sèvres near Paris. Its definition is simple:[16]

> 'The kilogram is the unit of mass; it is equal to the mass of the international prototype of the kilogram.'

The British Imperial units, like the yard and the pound, were defined similarly and standard prototypes were kept by the National Physical Laboratory in England and the National Bureau of Standards in Washington DC.

This trend for standardisation saw the creation of scientific units of measurement. As a result we habitually measure lengths, masses and times in multiples of metres, kilograms and seconds. One unit of each gives a familiar quantity that is easily imagined: a metre of cloth, a kilogram of potatoes. This convenience of size witnesses at once to their anthropocentric pedigree. But its inconvenience also becomes obvious when we start to use these units to describe quantities that are super- or sub-human in scale. The smallest atoms are 10 billion times smaller than a metre. The Sun is more than 10^{30} kilograms in mass. In Figure 2.1 we show the span of sizes and masses of significant objects in the Universe with ourselves added for perspective. We sit in between

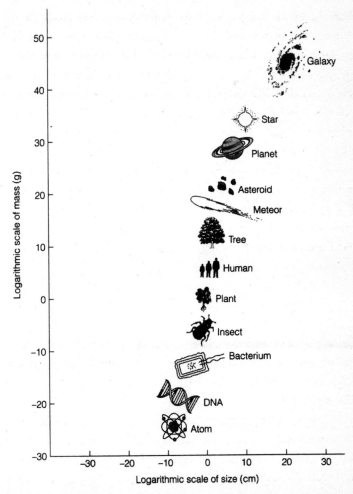

Figure 2.1 *The mass and size ranges of some important ingredients of the Universe. Our choice of centimetres and grams as units places us close to the centre of things.*

the huge astronomical distances and masses and the sub-atomic scale of the most elementary particles of matter.

Despite the introduction of universal metric standards by international commissions and government ministers, the ordinary worker took little notice of edicts about units, especially in Britain where a huge multiplicity of special units were in play throughout every branch of industry and commerce. By the middle of the nineteenth century, the industrial revolution had created diverse human sub-cultures of

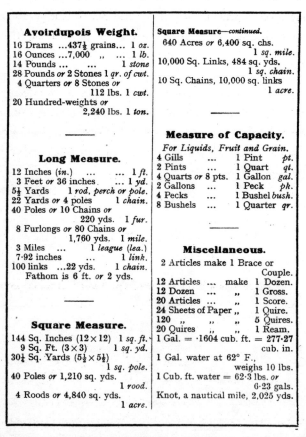

Avoirdupois Weight.
16 Drams ...437½ grains... 1 oz.
16 Ounces ...7,000 ,,   ... 1 lb.
14 Pounds ...   ...   1 stone
28 Pounds or 2 Stones 1 qr. of cwt.
 4 Quarters or 8 Stones or
                112 lbs. 1 cwt.
20 Hundred-weights or
             2,240 lbs. 1 ton.

Long Measure.
12 Inches (in.)   ...   ... 1 ft.
 3 Feet or 36 inches   ... 1 yd.
5½ Yards   1 rod, perch or pole.
22 Yards or 4 poles   1 chain.
40 Poles or 10 Chains or
                220 yds. 1 fur.
 8 Furlongs or 80 Chains or
              1,760 yds. 1 mile.
 3 Miles   ...   1 league (lea.)
7·92 inches   ...   1 link.
100 links ...22 yds. 1 chain.
   Fathom is 6 ft. or 2 yds.

Square Measure.
144 Sq. Inches (12 × 12) 1 sq. ft.
 9 Sq. Ft. (3 × 3)   1 sq. yd.
30¼ Sq. Yards (5½ × 5½)
                   1 sq. pole.
40 Poles or 1,210 sq. yds.
                   1 rood.
 4 Roods or 4,840 sq. yds.
                   1 acre.

Square Measure—continued.
 640 Acres or 6,400 sq. chs.
                   1 sq. mile.
10,000 Sq. Links, 484 sq. yds.
                   1 sq. chain.
10 Sq. Chains, 10,000 sq. links
                   1 acre.

Measure of Capacity.
For Liquids, Fruit and Grain.
4 Gills   ...   1 Pint   pt.
2 Pints   ...   1 Quart   qt.
4 Quarts or 8 pts. 1 Gallon gal.
2 Gallons   ...   1 Peck   pk.
4 Pecks   ...   1 Bushel bush.
8 Bushels   ...   1 Quarter qr.

Miscellaneous.
2 Articles make 1 Brace or
                   Couple.
12 Articles ... make 1 Dozen.
12 Dozen   ...   ,,   1 Gross.
20 Articles ...   ,,   1 Score.
24 Sheets of Paper ,,   1 Quire.
120   ,,   ,,   ,,   5 Quires.
20 Quires   ,,   ,,   1 Ream.
1 Gal. = ·1604 cub. ft. = 277·27
                   cub. in.
1 Gal. water at 62° F.,
                weighs 10 lbs.
1 Cub. ft. water = 62·3 lbs. or
                   6·23 gals.
Knot, a nautical mile, 2,025 yds.

Figure 2.2 *A typical set of miscellaneous weights and measures from an English self-help book of the 1950s.*[17]

engineers and brewers, accountants and metalworkers, timekeepers and ship workers, all of whom needed ways of measuring the materials that they managed and manipulated. The result was an explosion of units of measure. Every type of material began to have its own standard of strength and tolerance, quantity and weight. Not only were these units anthropocentric they were profession-centric as well. Brewers liked one choice of volume measure, water engineers another; jewellers measured weight differently to sailors and architects. When I was a child there was a common brand of lined exercise book that would be used for making notes at school. They always had red or blue covers and the outside back cover of the book listed all the peculiar Imperial measures of length, area, capacity and weight (see Figure 2.2).

For the engineer and the practical person of affairs this was convenient, useful and no doubt very profitable. But for anyone seeking an integrated natural philosophy it made human knowledge appear fragmented and idiosyncratic. A visitor from another planet would be perplexed by the need for different measures of weight when buying gold, apples or sealing wax.

MAINTAINING UNIVERSAL STANDARDS

'There was a crooked man who built a crooked house.'

Nursery rhyme

By the second half of the nineteenth century, engineers, industrialists and scientists were becoming overwhelmed by the profusion of *ad hoc* units and measures. The industrial revolution had accelerated the development of every imaginable industry. Manufacturing, machining, measuring, designing, building – these were the rages of the age and they spawned more and more units.

Within the halls of science the existence of standard lengths and masses was not entirely satisfactory for the purist either. Every time standard masses were handled with their special tongs their mass would be very slightly changed. It would vary slightly as atoms were evaporated from their surfaces or dust deposited from the atmosphere. They were not really constant.[18] Nor were they universal. Suppose that a signal had been received from an engineer on another planet asking us how big we were. It would be no use sending an answer in metres or kilograms and then responding to the inevitable reply, 'What are they?' by telling our extraterrestrial correspondent that they were objects kept in glass containers in Paris. Unfortunately the quest for universal standards had created examples which were neither standard nor universal.

Within science the driving force for rationalisation came from the study of electricity and magnetism. Different systems of units were in use by different groups of scientists and had different relationships with the traditional metric units for mass, length, time and temperature.

The first general response to these problems came from Lord Rayleigh and James Clerk Maxwell. In his Presidential address to the British Association for the Advancement of Science in 1870 Maxwell advocated the introduction of standards which are not tied to special objects, like standard metres[19] or kilograms kept in special conditions. For standards like these can never really be constant. The standard mass in Paris will lose and gain molecules all the time. Measures of time that are defined, like the day, by the rotation of the Earth or, like the year, by its orbit of the Sun likewise cannot be constant. As the rotation of the Earth slows, and our solar circuit changes, so these standards will very slowly drift. They may be defined in extrahuman terms but they are not candidates for ultimate standards. Maxwell had spent a good deal of time studying the behaviour of molecules in gases and was very impressed by the way in which each molecule of hydrogen was the same as all the others. This was quite different to dealing with large, everyday objects where every one was different. Maxwell saw an opportunity to use the sameness of molecules to define standards absolutely:

> 'Yet, after all, the dimensions of our earth and its time of rotation, though, relatively to our present means of comparison, very permanent, are not so by any physical necessity. The earth might contract by cooling, or it might be enlarged by a layer of meteorites falling on it, or its rate of revolution might slowly slacken, and yet it would continue to be as much a planet as before.
>
> But a molecule, say of hydrogen, if either its mass or its time of vibration were to be altered in the least, would no longer be a molecule of hydrogen.
>
> If, then, we wish to obtain standards of length, time, and mass which shall be absolutely permanent, we must seek them not in the dimensions, or the motion, or the mass of our planet, but in the wave-length, the period of vibration, and the absolute mass of these imperishable and unalterable and perfectly similar molecules [i.e. atoms].[20]

Maxwell was specially interested in molecules for many philosophical purposes. He recognised the significance of there existing populations of identical building blocks for all the material bodies we see around us. If we take any piece of pure iron it will be composed of a collection of identical iron molecules. The fact that these molecules appear to be identical is a remarkable feature of the world. Maxwell contrasted this invariance with the changeability and evolution of living things predicted by Charles Darwin's theory of evolution by natural selection. Maxwell pointed to the molecules of Nature as entities that were not subject to selection, adaptation or mutation. His challenge was to find a way to exploit this immutability and universality in the way that we define our units of measurement. In this way we would be able to take a step away from the bias introduced by the imperatives of human convenience towards the deep invariances of physical reality.

In 1905 the red light emitted by hot cadmium atoms[21] was first used as a standard against which to define a unit of length called the Ångstrom (denoted by 1Å and equal to 10^{-10} metre). One wavelength of the cadmium light was equal to 6438.4696 Å. This was a key step because for the first time it defined a standard of length in terms of a universally constant feature of Nature. The wavelength of the light emitted by cadmium[22] is fixed by the constants of Nature alone. If we wanted to tell an extraterrestrial physicist our size, we could do it by saying what we mean by 2.8 billion wavelengths of red cadmium light.[23]

A BRILLIANT IDEA!

'"Where did the matter come from?"

"What is the difference? . . .The secret of the universe is apathy. The earth, the sun, the rocks, they're all indifferent, and this is a kind of passive force. Perhaps indifference and gravitation are the same."'

Isaac Bashevis Singer[24]

In 1874, an unusual Irish physicist called George Johnstone Stoney found himself having to make sense of the Babel of practical units. He had been invited to deliver a lecture on units of measurement at the annual meeting of the British Association for the Advancement of Science in Belfast.[25] This annual meeting still exists today but is now devoted to showcasing the developments in science for the general public, the Press and young people. But in Stoney's day it was the foremost science conference in the world, a place where great discoveries would be made public and the Press would report on great debates between leading scientists and commentators. Today there are so many specialised scientific conferences, workshops, meetings, discussions, panel discussions and round tables that there is no longer any place for a meeting that covers all of science at a technical level – it would be impossibly big, impossibly lengthy, and well nigh unintelligible to most of the participants much of the time.

Stoney was an eccentric and original thinker. He was the first person to show how to deduce whether or not other planets in the solar system possessed a gaseous atmosphere, like the Earth, by calculating whether their surface gravity was strong enough to hold on to one. But his real passion was reserved for his most treasured idea – the 'electron'. Stoney had deduced that there must exist a basic ingredient of electric

Figure 2.3 *The Irish physicist George Johnstone Stoney (1826–1911).*[26]

charge. By studying Michael Faraday's experiments on electrolysis Stoney had even predicted[27] what its value must be – a prediction subsequently confirmed by J.J. Thomson who discovered the electron in Cambridge in 1897[28] and announced his discovery to the Royal Institution on 30 April. To this basic quota of electric charge Stoney eventually gave the name 'electron' and the symbol E in 1891[29] (after first calling it[30] the 'electrine' in 1874) and he never missed an opportunity to publicise its properties and potential benefits for science.[31]

Stoney was also an older distant cousin of the famous mathematician, computer scientist and code-breaker, Alan Turing, whose mother recalled childhood memories of the unusual uncle the children called 'electron Stoney'.[32] He was also the uncle of George FitzGerald, now famous for proposing the Lorentz-FitzGerald contraction of length, a phenomenon that was eventually understood within the context of Einstein's special theory of relativity. Stoney was also a practical man and worked for two years for the Earl of Rosse constructing sensitive optical instruments for his private observatory at Birr Castle before becoming Professor of Natural Philosophy at Queen's College Galway in 1850. After his retirement he moved to Hornsey in north London and continued publishing a steady stream of papers in the Royal Dublin Society's scientific journal. It is hard to find an issue that doesn't contain a paper under his name, on every conceivable subject, we find everything from time travel to how bicycles stay upright.

Stoney found the programme for the Belfast meeting of the British Association full of accounts of different units and standards: how to measure them; how best to define them; how to inter-relate them. This was all very useful for insiders but somewhat tedious for everyone else. Stoney saw an opportunity to simplify this vast perplexity of human standards of measurement and to do so in a way that would lend more weight to his electron hypothesis. Stoney had been a member of a British Association committee[33] which had determined conventions for electrical units in the years leading up to this conference, and so had already been required to give some thought to the problems of units and standards.

Stoney recognised that his concept of the basic electronic charge unit provided the missing piece in a small puzzle. Suppose that one wanted to devise units of mass, length and time that were not attached to purely human standards of convenience, like the pound or the mile or the fortnight. Then they needed to be derived from some aspect of the underlying fabric of the Universe that was not anthropocentric, that did not depend on where you are located when you measure it, or when the measurement was made. This ruled out traditional approaches to standards which took a standard mass of a kilogram or a length of a metre and kept them in a specially controlled environment somewhere and just compared other reference masses or lengths to them. These masses and lengths are anthropocentric in origin but, what is worse, they are anthropocentric in principle because there is no way in which to tell extraterrestrials how much mass or length defines our standard without sending it to them.

In order to escape the shackles of anthropocentric bias Stoney looked to the constants of physics to supply something that might transcend human standards of quantity. Newton had discovered that gravity obeys an apparently universal law. The force between two masses whose centres are separated by a distance is proportional to each of their masses and inversely proportional to the square of the distance between their centres. The constant of proportionality should be the same everywhere in the universe.[34] This constant, G, gives a measure of the strength of gravity. The important thing about it is that it is believed to be constant[35] – the same value should be found everywhere it is correctly measured. Moreover, it has a strange value when expressed in our convenient anthropocentric units ($G = 6.67259 \times 10^{-11} \, \mathrm{m^3 s^{-2} kg^{-1}}$) because those units were devised for other anthropocentric purposes.

The second constant of Nature that Stoney appealed to for his non-anthropocentric standards was the speed of light, c. Again, this quantity transcends human standards. It has a fundamental significance.

In fact, it has an even more fundamental significance than Stoney could ever have known. Einstein showed that the speed of light in a vacuum should act as the ultimate speed limit in the Universe – no information can be sent faster. It had also been discovered that the product of the permeability and permittivity of space that defined different units of electricity was equal to the square of the speed of light, so revealing its special universal status with respect to electricity as well. To these two constant quantities Stoney added his own candidate for the third great constant of Nature – his basic electron charge, which we now label by the symbol e. It was the last piece needed to complete the jigsaw. It fitted the bill in the same way as G and c. It was presumed to be universal. It was associated with a fundamental aspect of the structure of Nature. And it didn't care about human convenience. Stoney announced his trinity of constants like this:[36]

> 'Nature presents us with three such units and that if we take these as our fundamental units, instead of choosing them arbitrarily, we shall bring our quantitative expressions into a more convenient, and doubtless into a more intimate, relation with Nature as it actually exists.
>
> For such a purpose we must select phenomena that prevail throughout the whole of Nature, and are not simply associated with individual bodies. The first of Nature's quantities of absolute magnitude to which I will invite attention is that remarkable velocity of an absolute amount, independent of the units in which it is measured, which connects all systematic electrostatic units with the electromagnetic units of the same series. I shall call this velocity V_1 [i.e. our c]. If it were taken as our unit velocity we should at one stroke have an immense simplification introduced into our treatment of the whole range of electric phenomena, and probably into our study of light and heat.

Again Nature presents us with one particular coefficient of gravitation, of an absolute amount independent of the units in which it is measured, and which appears to extend to ponderable matter of every description throughout the whole material universe. This coefficient I shall call G_1 [i.e. our G]. If we were to take this as our unit of coefficients of attraction, it is presumable that we might thereby lay the foundation for detecting wherein lies the connection which we cannot but suspect between this most wonderful property common to all ponderable matter, and the other phenomena of nature.

And, finally, Nature presents us in the phenomenon of electrolysis, with a single definite quantity of electricity which is independent of the particular bodies acted on . . . This definite quantity of electricity I shall call E_1 [i.e. our e]. If we make this our unit quantity of electricity, we shall probably have made a very important step in our study of molecular phenomena.

Hence we have very good reason to suppose that in V_1, G_1, and E_1, [i.e. c, G and e] we have three of a series of systematic units that in an eminent sense are the units of Nature, and stand in an intimate relation with the work which goes on in her mighty laboratory.

We have thus obtained . . . the three great fundamental units offered to us by Nature, upon which may be built an entire series of physical units deserving of the title of a truly Natural Series of Physical Units.'

In his talk Stoney referred to the electron as the 'electrine' and gave the first calculation of its expected value.[37] He showed that the magic trio of G, c and e could be combined in one way, and only one way, so that a unit of mass, a unit of length and a unit of time are created from them. For the velocity of light he used an average of existing measurements, $c = 3 \times 10^8$ metres per second; for Newton's gravitation constant he used the value obtained by John Herschel, $G = 0.67 \times 10^{-11}$ m^3 $Kg^{-1}s^{-2}$, and for his unit of 'electrine' charge he used $e = 10^{-20}$ Ampères.[38] Here are the unusual new units that he found, in terms of the constants e, c and G, and in terms of grams, metres and seconds:

$$M_J = (e^2/G)^{1/2} = 10^{-7} \text{ gram}$$

$$L_J = (Ge^2/c^4)^{1/2} = 10^{-37} \text{ metres}$$

$$T_J = (Ge^2/c^6)^{1/2} = 3 \times 10^{-46} \text{ seconds}$$

These are extraordinary quantities. Although a mass of 10^{-7} gram is not too outlandish, similar to that of a speck of dust, Stoney's units of length and time were unlike any that had been encountered by scientists before. They were fantastically, almost inconceivably, small. There was (and still is) no possibility of measuring such lengths and times directly. In a way, that is what one might have expected. These units are deliberately not constructed from human dimensions, for human convenience, or for human utility. They are defined by the very fabric of physical reality that determines the nature of light, electricity and gravity. They don't care about us.

Stoney had succeeded brilliantly in his quest for a superhuman system of units. But, alas, they attracted little attention. There was no practical use for his 'natural' units and their significance was hidden to everyone, even Stoney himself, who was more interested in promoting his electron up until its discovery in 1897. Natural units needed to be discovered all over again.

MAX PLANCK'S NATURAL UNITS

'Science cannot solve the ultimate mystery of nature. And that is because, in the last analysis, we ourselves are part of the mystery that we are trying to solve.'

Max Planck[39]

Stoney's idea was rediscovered in a slightly different form by the German physicist Max Planck, in 1899. Planck is one of the most important physicists of all time. He discovered the quantum nature of energy that launched the quantum revolution in our understanding of the world and provided the first correct description of heat radiation (the so called 'Planck spectrum') and has one of the fundamental constants of Nature named after him. He was a central figure in physics of his time, won the Nobel prize for physics in 1918, and died in 1947 aged 89. A quiet, unassuming man, he was deeply religious[40] and greatly admired by his younger contemporaries, like Einstein and Bohr.

Planck's conception of Nature placed great emphasis upon its intrinsic rationality and independence of human thought. He believed in an intelligence behind the appearances which fixed the nature of reality. Our most fundamental conceptions of Nature needed to be aware of the need to identify that deep structure which was far from the needs of human utility and convenience. In the last year of his life he was asked by a former student if he believed that the quest to unite all the constants of Nature by some deeper theory was appealing. He replied with enthusiasm, tempered by realism about the difficulty of the challenge:

'As to your question about the connections between the universal constants, it is without doubt an attractive idea to link them together as closely as possible by reducing these various constants to a single one. I for my part, however, am doubtful that this will be successful. But I may be mistaken.'[41]

Unlike Einstein, Planck did not really believe in any attainable all-encompassing theory of physics which would explain all the constants of Nature. For if such a theory arrived then physics would cease to be an inductive science. Others, like Pierre Duhem and Percy Bridgman, regarded the promised Planckian separation of scientific description from human conventions as unattainable in principle, viewing the constants of Nature and the theoretical descriptions that they underpin entirely as artefacts of a particular human choice of representation to make sense of what was seen.

Planck was suspicious of attributing fundamental significance to quantities that had been created as a result of the 'accident' of our situation:

> 'All the systems of units that have hitherto been employed, including the so-called absolute C. G. S. system [centimetre, gram and second, for measuring length, mass and time], owe their origin to the coincidence of accidental circumstances, inasmuch as the choice of the units lying at the base of every system has been made, not according to general points of view which would necessarily retain their importance for all places and all times, but essentially with reference to the special needs of our terrestrial civilization . . .
>
> Thus the units of length and time were derived from the present dimensions and motion of our planet, and the units of mass temperature from the density and the most important temperature points of water, as being the liquid which plays the most important part on the surface of the earth, under a pressure which corresponds to the mean properties of the atmosphere surrounding us. It would be

no less arbitrary if, let us say, the invariable wave length of Na-light were taken as unit of length. For, again, the particular choice of Na from among the many chemical elements could be justified only, perhaps, by its common occurrence on the earth, or by its double line, which is in the range of our vision, but is by no means the only one of its kind. Hence it is quite conceivable that at some other time, under changed external conditions, every one of the systems of units which have so far been adopted for use might lose, in part or wholly, its original natural significance.'

Instead, he wanted to see the establishment of

'units of length, mass, time and temperature which are independent of special bodies or substances, which necessarily retain their significance for all times and for all environments, terrestrial and human or otherwise'.[42]

Whereas Stoney had seen a way of cutting the Gordian knot of subjectivity in the choice of practical units, Planck used his special units to underpin a non-anthropomorphic basis for physics and 'which may, therefore, be described as "natural units".' The progressive revelation of this basis was for him the hallmark of real progress towards as far-reaching a separation as possible of the phenomena in the external world from those in human consciousness.

In accord with his universal outlook, in 1899 Planck proposed[43] that natural units of mass, length and time be constructed from the most fundamental constants of Nature: the gravitation constant G, the speed of light c, and the constant of action, h, which now bears Planck's name.[44] Planck's constant determines the smallest amount by which energy can be changed (the 'quantum'). In addition, the incorporation of Boltzmann's constant, k – which simply converts units of energy into units of temperature — allowed him also to define a natural temperature.[45] Planck's units are the only combinations of these constants which can be formed with the dimensions of mass, length, time and temperature. Their values are not very different from Stoney's:

$$m_{pl} = (hc/G)^{1/2} = 5.56 \times 10^{-5} \text{ gram}$$

$$l_{pl} = (Gh/c^3)^{1/2} = 4.13 \times 10^{-33} \text{ centimetres}$$

$$t_{pl} = (Gh/c^5)^{1/2} = 1.38 \times 10^{-43} \text{ seconds}$$

$$T_{pl} = k^{-1}(hc^5/G)^{1/2} = 3.5 \times 10^{32} \text{ Kelvin}$$

Again, we see a contrast between the small, but not outrageously small natural unit of mass and the fantastically extreme natural units of time, length and temperature.[46] These quantities had a superhuman significance for Planck. They cut into the bedrock of physical reality:

> 'These quantities retain their natural significance as long as the law of gravitation and that of the propagation of light in a vacuum and the two principles of thermodynamics remain valid; they therefore must be found always to be the same, when measured by the most widely differing intelligences according to the most widely differing methods.'

He alludes in his closing words to the idea of observers elsewhere in the Universe defining and appreciating these quantities in the same way as ourselves.[47]

For the time, there was something quite striking about Planck's units, as there was also about Stoney's. They entwined gravity with the constants governing electricity and magnetism. Gravity had always been a largely uneventful branch of physics. Newton had apparently found the law of gravity and very few questions were asked about it thereafter. True, there were annoying small discrepancies between its predictions and the observed wobble of the planet Mercury as it orbited close to the Sun. Some had even suggested making a tiny change to Newton's law to explain it but most astronomers expected that small effects from the non-spherical shape of the Sun or errors in the observations might rescue Newton. It seemed to be a finished story.

By contrast, there was continual progress and debate about the laws of electricity and magnetism. They began looking like separate laws for static electricity (that makes your hair stand on end), dynamic electricity (that makes currents flow), and magnetism. But gradually the two electricities were found to be different complexions of one electric force. And then Maxwell showed that electricity and magnetism were really different sides of the same coin: moving magnets could make electrical currents flow and electric currents could create magnetic forces. But never did gravity seem to impinge upon electricity and magnetism or the behaviour of atoms and molecules. As a result we see that there existed a very different view to that of Planck and Stoney about natural units. The physicist Paul Drude, a leading contributor to the study of electromagnetic waves, optics and materials, held the prestigious professorship of physics at Leipzig. In 1897 Drude proposed[48] a system of absolute units of mass, length and time that were tied to the properties of the aether that was then believed to permeate all space. His choice of standards were the velocity of light, and the average distance travelled by the particles of the aether before they interacted. Drude could then see no way[49] for gravity to be linked to electricity and magnetism and so did not follow Stoney and Planck in devising natural units containing G. Even for Planck, the entry of G into his natural units was a mystery. He offered no explanation as to the meaning of the tiny Planck units of length and time. What did they mean? What would happen if you looked at the world on these dimensions? It would be a long time before these questions were asked[50] and far longer before they were answered.

PLANCK GETS REAL

'The increasing distance of the physical world picture from the world of the senses means nothing but a progressive approach to the real world.'

Max Planck

We have seen how Max Planck appealed to the existence of universal constants of Nature as evidence for the existence of a physical reality that was quite distinct from human minds. But he wanted to go much further and use the existence of these immutable constants as an argument against positivistic philosophers who thought science was entirely a human edifice: measured points organised in a convenient way by a theory that will eventually be replaced by a better one. Planck appreciated that the writing of equations and the formulation of physical theories was a human activity, but that does not mean that it is nothing but a human activity. For him, the constants of Nature had emerged uninvited and, as his natural units clearly showed, were not chosen for human convenience alone. He writes:[51]

> 'These . . . numbers, the so-called "universal constants" are in a sense the immutable building blocks of the edifice of theoretical physics.
>
> So now we must continue with the question: What is the real meaning of these constants? Are they, in the last analysis, inventions of the inquiring mind of man, or do they possess a real meaning independent of human intelligence?
>
> The first of these two views is professed by the followers of positivism, or at least by its most extreme partisans. Their theory is that physical science has no other foundation than the measurements on which its structure

is erected, and that a proposition in physics makes any sense only in so far as it can be supported by measurements.

Therefore, up to quite recently, positivists of all hues have also put up the strongest resistance to the introduction of atomic hypotheses and thereby also to the acceptance of the above mentioned universal constants. This is quite understandable, for the existence of these constants is a palpable proof of the existence in nature of something real and independent of every human measurement.

Of course, even today a consistent positivist could call the universal constants mere inventions which have proved to be uncommonly useful in making possible an accurate and complete description of the most diversified results of measurement. But hardly any real physicist would take such an assertion seriously. The universal constants were not invented for reasons of practical convenience, but have forced themselves upon us irresistibly because of the agreement between the results of all relevant measurements, and – this is the essential thing – we know quite well in advance that all future measurements will lead to these selfsame constants.'

There were many more options open to Planck's opponents, of course. It might have been that the constants he chose were not truly constants at all when scrutinised with vastly greater precision. They might be varying very slowly, perhaps by only a few parts per million over the age of the Universe. Or, it might be that they are only constant in some statistical or average sense. Since these possibilities cannot be excluded except by assumption or prejudice there needs to be a detailed experimental study of constants and their constancy. Physicists became interested in determining the values of the constants of Nature with greater and greater accuracy and devising ways of checking whether they were truly constant. This quest for the evaluation of the constants of Nature had seemed to some to be the ultimate goal of physics. For, amusingly, at the end of the nineteenth century, it was widely believed that all the interesting discoveries had already been made in physics and all that remained was to measure with greater and greater accuracy – an enterprise of polishing rather than of discovery or revolution. Caricaturing this hubris, Albert Michelson wrote in 1894 that there was a view abroad that

> 'The more important fundamental laws and facts of physical science have all been discovered, and these are now so firmly established that the possibility of their ever being supplanted in consequence of new discoveries is remote . . . Our future discoveries must be looked for in the sixth place of decimals.'[52]

Even Planck had been influenced by these views. As a student in 1875 he recalled that his tutor advised him to work in biology because all the important problems of physics were solved and the subject was fast approaching completeness. Ironically, Planck was the leader in creating the new quantum view of reality that was then followed by Einstein's assaults on our conceptions of space, time and gravity. Far from being near to completion, physics had barely begun.

ABOUT TIME

> 'The old believe everything: the middle-aged suspect everything: the young know everything.'
>
> Oscar Wilde[53]

One of the paradoxes of our study of the Universe around us is that as our descriptions of its workings become more precise and successful so they also become increasingly remote from everyday human experiences. The most accurate predictions that we can make are not about the workings of banks or the vagaries of consumer choice and voter intent, they are about elementary particles and astronomical systems of spinning stars. This is exactly the opposite to what would be expected if our descriptions of the world were strongly biased by input from the human mind rather than being in some sense acts of discovery. It need not have been like this. We have only to look at our attempts to understand the complexities of human behaviour to recognise a strong subjective element. The reliability of our conclusions generally falls as we deal with situations farthest from our own experience and individuals least like ourselves.

By contrast, our unravelling the existence of constants of Nature behind the realities described by laws of change and invariance has enabled us to formulate standards by which we can judge whether things are big or small, young or old, heavy or light, hot or cold, by reference to an absolute standard. When we say that the Universe has been expanding for 13 billion years, does that mean it is *old*? It sounds very old against the fleeting span of a human lifetime, or when compared with the day or the year that derive from the motions of the Earth. But, then again, the Universe might be going to expand for trillions of years, or perhaps even forever. By those standards it is very young. Natural units tell us that in a well-defined sense the Universe *is* very old already, about 10^{60} Planck times old. Life on Earth didn't appear until after the Universe was 10^{59} Planck times old. We were a late arrival.

chapter three

Superhuman Standards

'Brother Mycroft is coming round.'

A. Conan Doyle[1]

EINSTEIN ON CONSTANTS

'What I'm really interested in is whether God could have made the world in a different way; that is, whether the necessity of logical simplicity leaves any freedom at all.'

Albert Einstein[2]

Albert Einstein did more than any other scientist to create the modern picture of the laws of Nature. He played a major role in creating the correct perspective upon the atomic and quantum character of the small-scale world of matter, showed how the speed of light introduced a relativity into each observer's view of space, mass and time, and single-handedly found the theory of gravity that superseded the classic picture created by Isaac Newton 250 years before. He was always fascinated by the way in which some things must always look the same, no matter how the viewer is moving. The prime example that he displayed was the speed of light moving in a vacuum. No matter how fast the source of a light beam is moving relative to you, after it emits its light you will always measure the light to have the same speed relative to you.

This is completely unlike any everyday motion at low speed that we are familiar with. Launch a missile at 500 kilometres per hour from a train that is moving in the same direction at 100 kilometres per hour and the missile will be found to move at 600 kilometres per hour relative to the ground. But fire a light beam from a train moving at the speed of light (300,000 kilometres per second) and it will be found to move at the speed of light relative to the ground. The speed of light is a special constant of Nature. It is the benchmark against which we can judge whether motion is 'fast' or 'slow' in some absolute sense. All over the Universe we expect that the speed of light plays the same basic role. It is a cosmic speed limit: no information can be transferred faster than the speed of light in vacuum.[3]

Einstein had many interesting things to say about the constants of Nature at different stages of his life. It was his elucidation of the theory of relativity that endowed the velocity of light in vacuum with its special status as the maximum speed at which information could be transmitted in the Universe. He revealed the full extent of what Planck and Stoney had merely assumed: that the velocity of light was one of the fundamental superhuman constants of Nature. In the second half of his life, he became increasingly absorbed with a search for the ultimate theory of physics. He called it a 'unified field' theory whereas today it would be called a 'Theory of Everything'.[4] Alas, physicists now believe that Einstein achieved very little in that period of intense investigation, as he constantly tried to find a bigger and better theory than his general theory of relativity: one that would include other forces of Nature than gravity.[5] He believed that such a theory existed and its uniqueness and completeness would leave no mathematical loose ends. Consequently, it would have the smallest possible number of constants of Nature[6] which could then only be found by experiment.

Einstein was not really happy for there to be any free constants like this at all. He realised that the search for the ultimate theory was a process of finding better and better theories which superseded the previous one. At present our theories are provisional and so there are a number of free constants of Nature appearing in them which we just have to measure. Ultimately, this situation would change. He expected that his unified theory would determine the values of constants like e, G and c in terms of pure numbers that could be calculated as accurately as one wished.

Einstein wrote almost nothing about these ideas in his published articles and other scientific writings. Yet he maintained a lifelong correspondence with an old student friend, Ilse Rosenthal-Schneider (pictured in Figure 3.1), who was interested in the philosophy of science and was a close friend of both Planck and Einstein in her youth. She and her husband emigrated to Sydney to escape from Nazi Germany in 1938. For a period, between 1945 and 1949, the personal letters between Einstein and Rosenthal-Schneider focused on the question of constants of Nature. Einstein thinks carefully about his explanations and provides a clear and full statement of his beliefs and hopes for the future of physics.

Rosenthal-Schneider first wrote[7] to Einstein about constants in 1945. What are they? What are they telling us about the lawfulness of Nature? Are they all related? She was surprised to get a very fast response which actually began to answer her questions. She had learned that questions about his health, general situation or other personal matters generally went unanswered or were ignored in his replies. But this was a subject he wanted to think about. His reply was posted from Princeton on 11 May 1945:

> 'With the question of the universal constants, you have broached one of the most interesting questions that may be asked at all. There are two kinds of constants: apparent and real ones. The apparent ones are simply the outcome of the introduction of arbitrary units, but are eliminable. The real [true] ones are genuine numbers which God had to choose arbitrarily, as it were, when He deigned to create this world.

Figure 3.1 *Ilse Rosenthal-Schneider (1891–1990).*[8]

My opinion now is – stated briefly – that constants of the second type do not exist and that their apparent existence is caused by the fact that we have not penetrated deeply enough. I therefore believe that such numbers can only be of a basic type, as for instance π or e.'

What Einstein is saying is that there are some apparent constants which are created by our habit of measuring things in particular units. The radiation constant of Boltzmann's is like this. It is just a conversion factor between energy and temperature units, rather like the conversion factors between Fahrenheit and centigrade scales of temperature. The true constants have to be pure numbers, not quantities that have 'dimensions', like a speed or a mass or a length. Quantities with dimensions always change their numerical values if we change the units in which they are expressed. Even the speed of light in vacuum can't be one of the true constants Einstein is searching for. A speed has units of length per unit time and so could not be shown to be some combination of the 'basic' numbers, like π, that

Einstein seeks. It could equally be 186,000 miles per second or 300,000 kilometres per second. These two numbers can't be explained by an ultimate theory of physics. Instead, we must find another constant of Nature which has the dimensions of a velocity. The ratio of this quantity to the speed of light will then be a pure number, with no dimensions. There is then the possibility that it might be a number that could be calculated in terms of quantities like π or any of the other numbers of mathematics.

Rosenthal-Schneider replies[9] and mentions the ideas of Planck, with whom she studied as a student, about the three special constants that he used to create his 'natural' units:

> 'However, I am still worrying – and that is why I pester you again with my questions – about what are the universal constants as Planck used to enumerate them: gravitational constant, velocity of light, quantum of action, . . . which are not dependent on external conditions like pressure, temperature, . . . and which therefore are pleasantly distinct from the constants of irreversible processes? If all these were entirely non-existent, the consequences would be catastrophic,
>
> If I understood Planck correctly he regarded such universal constants as "absolute quantities." If now you were to state that they are all non-existent, what at all would be left for us in the natural sciences? It is much more worrying for an ordinary mortal than you can imagine.'

Einstein's penfriend is worried about the consequences of there being no true constants of Nature. If they are all illusory, what bedrock is there for physical reality; why does the Universe seem to be the same from one day to the next? She misunderstands Einstein's statement that there are no free constants of Nature, thinking that he means that they are not constant when he means only that he believed they are not free.

A deeper theory will eventually determine them. Sensing that he has misled his correspondent, he responds in greater detail[10] on 13 October 1945, with a complete analysis of the situation. First, he notes that there are just quantities like 2, π or e (a numerical constant equal to about 2.718) which appear in physical formulae. In a later chapter we shall discuss them further. Einstein notices that they tend to appear in physical formulae but their values are neither very large nor very small:[11] they are never very different from the number 1. They might be ten times greater or smaller but not millions of times greater or smaller. This is something he cannot explain. It just seems like a piece of good luck for physicists.[12]

> 'I see from your letter that you did not grasp my hint about the universal constants of physics. I will therefore try to make the matter clearer.
>
> 1. Basic numbers. These are those which, in the logical development of mathematics, appear by a certain necessity as unique individual formations.
>
> e.g., $e = 1 + 1 + 1/2! + 1/3! + \ldots$
>
> It is the same with π, which is closely connected with e. In contrast to such basic numbers are the remaining numbers which are not derived from 1 by means of a perspicuous construction.
>
> It would seem to lie in the nature of things that such basic numbers do not differ from the number 1 in respect of the order of magnitude, at least as long as consideration is confined to "simple" or, as the case may be, "natural" formations. This proposition, however, is not fundamental and not sharply definable.'

But Einstein knows that these basic numbers are not the most interesting constants of Nature. Einstein explains that the usual constants, like the speed of light, Planck's constant, or the gravitation constant, have dimensions of different powers of mass, length and time. From them we can create combinations which are pure numbers but we might need to introduce other quantities to do it. He says,

> 'Now let there be a complete theory of physics in whose fundamental equations the "universal" constants $c_1, \ldots c_n$ occur. The quantities may somehow be reduced to gm. cm. sec. The choice of these three units is obviously quite conventional. Each of these $c_1, \ldots c_n$ has a dimension in these units. We now will choose conditions in such a way that c_1, c_2, c_3 have such dimensions that it is not possible to construct from them a dimensionless product $c^\alpha_1 c^\beta_2 c^\gamma_3$. Then one can multiply $c_4 c_5$, etc., in such a way by factors built from powers of c_1, c_2, c_3 that these new symbols, c^*_4, c^*_5, c^*_6 are pure numbers. These are the genuine universal constants of the theoretical system which have nothing to do with conventional units.

Suppose his c_1, c_2, c_3 are Planck's c, h and G, then there is no way to combine them in powers so that you can get a pure number with no dimensions.[13] To do that you need to multiply by some other dimensional constants of Nature. For example, by multiplying G/hc by the square of some mass, for example the mass of a proton, we get the pure number Gm_{pr}^2/hc, say c^*_4, which is approximately equal[14] to 10^{-38}. The 'starred' number we have just created is made by measuring some constant of nature with units of a mass by Planck's mass. We could make others by dividing some time by Planck's time or some length by Planck's length. It is these pure 'starred' numbers that Einstein regards as the most fundamental. It does not matter what units are employed to measure them or to express them, they will always have the same value. Where do they come from? What fixes them? Why is Gm_{pr}^2/hc about equal to 10^{-38}, rather than to 10^3 or 10^{-68}? Einstein doesn't know, but he has a strong belief that they are fixed absolutely.[15] There is no latitude for them to be different:

'My expectation now is that these constants c^*_4 etc., must be basic numbers whose values are established through the logical foundation of the whole theory.

Or one could put it like this: In a reasonable theory there are no dimensionless numbers whose values are only empirically determinable.

Of course, I cannot prove this. But I cannot imagine a unified and reasonable theory which explicitly contains a number which the whim of the Creator might just as well have chosen differently, whereby a qualitatively different lawfulness of the world would have resulted.

Or one could put it like this: A theory which in its fundamental equations explicitly contains a non-basic constant would have to be somehow constructed from bits and pieces which are logically independent of each other; but I am confident that this world is not such that so ugly a construction is needed for its theoretical comprehension.'

Elsewhere, Einstein, is famously quoted as saying what really interests him is whether 'God had any choice in making the world'. What he meant by that cryptic statement is made clear by the extract from his letter to Rosenthal-Schneider. He wants to know whether the dimensionless constants of Nature could have been given different numerical values without changing the laws of physics or whether there is only one possible choice for them. Going further he might wonder whether different choices in their values are possible for different laws of Nature. We still don't know.[16]

The illuminating exchange of letters with Rosenthal-Schneider on constants ends on 24 March 1950 with Einstein reiterating his 'religious' view that God did not have any choice when it came to the fundamental constants and their values:

'Dimensionless constants in the laws of nature, which from the purely logical point of view can just as well have different values, should not exist. To me, with my "trust in God" this appears to be evident, but there will be few who are of the same opinion.'

As we leave Einstein's thoughts about the inevitability of the constants of Nature it is interesting to pick up on the view of other great physicists who have been drawn to speculate about the significance and attainability of a final understanding of their values. Take George Gamow, the eccentric Russian physicist who risked his life escaping from the Soviet Union to live and work in America, where he became one of the founders of modern cosmology and even contributed to the early understanding of the DNA molecule and the genetic code. Gamow, like all his contemporaries, could see that there were four distinct forces of Nature (gravity, electromagnetism, weak and strong forces). The strength of each would create one of Einstein's pure numbers that define the world. Gamow was not drawn especially into the issue of whether they could have only one quartet of possible values. But for him a full understanding of those values – an ability to calculate or predict them precisely – would be like the waving of the chequered flag to a physicist. They would have attained a complete understanding of the forces of Nature when that day happened. Gamow is a little depressed at the prospect, like reaching the end of a great story, or sitting at the summit of a mountain one has striven to scale, for

'If and when all the laws governing physical phenomena are finally discovered and all the empirical constants occurring in these laws are finally expressed through the four independent basic constants, we will be able to say that physical science has reached its end, that no excitement is left in further explorations, and that all that remains to a physicist is either tedious work on minor details of the self-educational study and adoration of the magnificence of the completed system. At that stage physical science will enter from the epoch of Columbus and Magellan into the epoch of *National Geographic Magazine*.'[17]

THE DEEPER SIGNIFICANCE OF STONEY-PLANCK UNITS: THE NEW *MAPPA MUNDI*

'One Ring to rule them all, One Ring to find them.
One Ring to bring them all and in the darkness bind them.'

J.R.R. Tolkien[18]

The interpretation of the natural units of Stoney and Planck was not at all obvious to physicists. Aside from occasional passing remarks it was not until the late 1960s that the renewed study of cosmology led to a full appreciation of these strange standards. One of the curious problems of physics is that it has two beautifully effective theories – quantum mechanics and general relativity – but they govern different realms of Nature.

Quantum mechanics holds sway in the microworld of atoms and elementary particles. It teaches us that every mass in Nature, however solid or pointlike it may appear, has a wavelike aspect. This wave is not like a water wave. It is more analogous to a crime wave or a wave of hysteria: it is a wave of information. It tells you the probability that you will detect a particle. If an electron wave passes through your detector you will be more likely to make a detection, just as you are more likely to be robbed if a crime wave hits your neighbourhood. The quantum wavelength of a particle gets smaller the more massive the particle. Situations are dominated by quantum waviness when the quantum wavelength of their participants exceeds their physical size. Everyday objects, like cars and speeding cricket balls, have such high masses that their quantum wavelengths are vastly smaller than their sizes and we can forget about quantum influences when driving cars or watching cricket matches.

By contrast, general relativity was always necessary when dealing with situations where anything travelled at a speed at, or close to, the speed of light or where gravity is very strong. It is used to describe the expansion of the universe and the behaviour of extreme situations like the formation of black holes. However, gravity is very weak compared with the forces that bind atoms and molecules together and far too weak to have any effect on the structure of atoms or sub-atomic particles.

As a result of these properties, quantum theory and gravitation govern different kingdoms that have little cause to talk to one another. This is fortunate. No one knows how to join the two theories together seamlessly to form a new, bigger and better edition that could deal with quantum aspects of gravity. All the candidates remain untested. But how can we tell when such a theory is essential? What are the limits of quantum theory and Einstein's general relativity theory? Fortunately, there is a simple answer and Planck's units tell us what it is.

Suppose we take the whole mass inside the visible Universe[19] and determine its quantum wavelength. We can ask when this quantum wavelength of the visible Universe exceeds its size. The answer is when the Universe is smaller than the Planck length in size (10^{-33} cm), less than the Planck time in age (10^{-43} secs), and hotter than the Planck temperature (10^{32} degrees). Planck's units mark the boundary of applicability of our current theories. To understand what the world is like on a scale smaller than the Planck length we have to understand fully how quantum uncertainty becomes entangled with gravity. To understand what might have gone on close to the event that we are tempted to call the beginning of the Universe or the beginning of time we have to penetrate the Planck barrier. The constants of Nature mark out the frontiers of our existing knowledge and show us where our theories start to overreach themselves.

In the recent attempts that have been made to create a new theory to describe the quantum nature of gravity a new significance has emerged for Planck's natural units. It appears that the concept we call 'information' has a deep significance in the universe. We are used to living in what is sometimes called 'the information age'. Information can be packaged in more electronic forms, dispatched more quickly and received more easily than ever before. The progress we have made in processing information quickly and cheaply is commonly displayed in a form which enables us to check the prediction of Gordon Moore, the founder of Intel, called Moore's Law (see Figure 3.2). In 1965, Moore noticed that the area of a transistor was being halved approximately every 12 months. In 1975 he revised this halving time to 24 months. This is 'Moore's Law': that every 24 months you get about twice as much computer circuitry, running at twice the speed, for the same price because the cost of integrated circuit remains roughly constant.

The ultimate limits that we can expect to be placed upon information storage and processing rates are imposed by the constants of Nature. In 1981, an Israeli physicist, Jacob Bekenstein, made an unusual prediction that was inspired by what he knew from the study of black holes. He calculated that there is a maximum amount of information that can be stored inside any volume. This should not surprise us. What should is that the maximum value is just determined by the surface area surrounding the volume, not the volume itself. The maximum number of bits of information that can be stored in a volume is just given by computing its surface area in Planck units. Suppose that the region is spherical. Then its surface area is just proportional to the square of its radius, while the Planck area is proportional to the Planck

Figure 3.2 *Moore's Law shows the evolution of computer processing speed versus time. Every two years the number of transistors that can be packed into a given area of integrated circuit doubles. This biennial halving of transistor size means that the computing speed of each transistor doubles every two years for the same cost.*

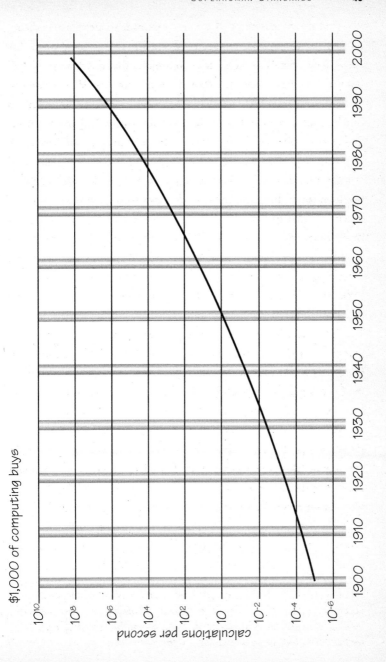

$1,000 of computing buys

length squared (10^{-66} cm^2). The total number of bits in a sphere of radius R centimetres is therefore just given by $10^{66} \times R^2$. This is vastly bigger than any information storage capacity that has so far been produced. Similarly, there is an ultimate limit on the rate of processing of information that is imposed by the constants of Nature.

It is also very remarkable that we are able to use the Planck and Stoney units to classify the whole range of structures that we see in the Universe, from the world of elementary particles up to the largest astronomical structures. These are shown in Figure 3.3. The structures shown on this picture are the stable entities that exist in the Universe. They exist because they are stable balancing acts between competing forces of attraction and repulsion. For example, in the case of a planet, like Earth, an equilibrium arises between the attractive crushing force of gravity and the atomic repulsion that arises when atoms are squeezed too close together. All these equilibria can be roughly expressed in terms of two pure numbers created from the constants e, h, c, G and m_{pr}

$$\alpha = 2\pi e^2/hc \approx 1/137 \text{ and } \alpha_G = Gm_{pr}^2/hc \approx 10^{-38}$$

There are three interesting things to say about this picture. First we notice that most things lie along a straight line running diagonally upwards from left to right. This line corresponds to the track of constant density that is equal to what we call 'atomic density'. Everything that is made of atoms has a density quite close to the density of a single atom given by the mass of an atom divided by its volume.[20] Second, there are some large empty areas in this picture. If we add to the picture the line defining where black holes and their interior regions sit then we take out the whole of the top left triangle of the picture. Nothing from this region could be seen. Its gravity would be too strong to allow light to escape. Similarly, nothing in the bottom left corner triangle would be detectable. This 'quantum region' contains objects which are so small that the act of observing them would perturb them into another part of the picture.

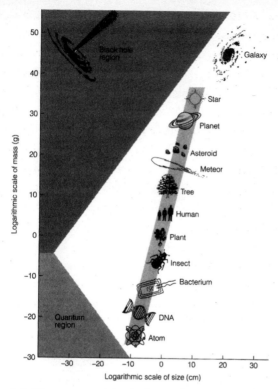

Figure 3.3 *The pattern with which the observed structures in the Universe fill the size-mass plane is dictated by three factors. The line of constant atomic density, the line that marks the black hole region within which things would be inside black holes (and hence invisible), and the line marking the uncertainty principle of quantum mechanics which separates off the quantum region in which the usual concepts of size and mass cannot be simultaneously maintained. We see that most of the familiar structures in the universe lie on or close to the line of constant atomic density. Along this line the mass of solid objects is proportional to their volume, or roughly to the cube of their sizes.*[21]

This is the region that is guarded by Heisenberg's Uncertainty Principle. Nothing within it is observable. However, it tells us our third interesting thing. Notice that the quantum line intersects the black hole line. This is the place where gravity and quantum reality collide. And what is that point: it has the Planck mass and the Planck size. Planck's units are the fulcrum around which the scales of reality turn.

OTHERWORLDLINESS

'Why did George Best leave the Barcelona–Manchester United match five minutes before the end? Because he was videoing the game and didn't want to know the result.'

Angus Deayton[22]

The identification of dimensionless constants of Nature like α and α_G, along with the numbers that play the same defining role for the weak and strong forces of Nature encourages us to think for a moment about worlds other than our own. These other worlds may be defined by laws of Nature which are the same as those which govern the Universe as we know it but they will be characterised by different values of dimensionless constants. These numerical shifts will alter the whole fabric of the imaginary worlds. The balances between their forces will be different from those in our world. Atoms may have different properties. Gravity may play a role in the small-scale world. The quantum nature of reality may enter in unexpected places.

The legitimacy of this little thought experiment is closely linked to Einstein's deep questions. If the laws of Nature allow one and only one set of the values for the constants of Nature then the freedom we appear to have to consider worlds in which they are different is merely a consequence of our relative ignorance. We think there is freedom to change their values without constraint only because we do not understand the extent to which those values are hardwired into the forms of the laws themselves. On the other hand, if the constants are not uniquely fixed by the one and only possible form for the laws of Nature then there may exist other worlds where they take different values.

The last important lesson we learn from the way that pure numbers like α define the world is what it really means for worlds to be different. The pure number that we call the fine structure constant and denote by α is a combination of the electron charge, e, the speed of light, c and Planck's constant, h. At first we might be tempted to think that a world in which the speed of light was slower would be a different world. But this would be a mistake. If c, h and e were all changed so that the values they have in metric (or any other) units were different when we looked them up in our tables of physical constants, but the value of α remained the same, this new world would be *observationally indistinguishable* from our world. The only thing that counts in the definition of the world are the values of the dimensionless constants of Nature. If all masses are doubled in value you cannot tell because all the pure numbers defined by the ratios of any pair of masses are unchanged.

THE SUPER-COPERNICAN PRINCIPLE

'A physicist is a mathematician with a feeling for reality.'

Norman Packard[23]

The name of the great Polish astronomer, Nicolaus Copernicus, is linked forever with the move to relinquish the presumption that the Earth is at the centre of things. For Copernicus himself it was the assumption, held for thousands of years, that the Earth was at the centre of the solar system that was the focus of attention. Copernicus constructed a picture of the motions of the planets around the Sun in which the Earth was no longer central. In time this heliocentric model proved the superior description of what was seen by astronomers, surpassing the explanatory power of the ancient Earth-centred picture of Ptolemy and his successors.

The impact over the following centuries of Copernicus' leap away from the prejudices of anthropocentrism was felt across the whole spectrum of human investigation. We began to appreciate our place in the Universe was by no means central. Indeed, in many respects, it appeared to be almost peripheral.

The march towards established constants of Nature that were not explicitly anthropocentric, but based upon the discovery and definition of universal attributes of Nature, can be seen as a second Copernican step. The fabric of the Universe and the pivotal structure of her universal laws were now seen to flow from standards and invariants that were truly superhuman and extraterrestrial. The fundamental standard of time in Nature bore no simple relation to the ages of man and woman, no link to the periods of days, months and years that defined our calendars, and was too short to allow any possibility of direct measurement.

There was a third step still to be taken in this extension of the Copernican perspective. It was to show that the *laws* of Nature displayed a Copernican complexion. This is a much subtler matter and required one of Einstein's greatest insights to carry it out. First, what does it mean?

Einstein argued that the laws of Nature should appear to be the same for all observers in the Universe, no matter where they were or how they were moving. If they were not then there would exist privileged observers for whom the laws of Nature looked simpler than they did for other observers. Such a view would be anti-Copernican; it would give someone (not necessarily us on Earth) a special position in the Universe. At first one might think that having universal constants of Nature based on superhuman physical standards would be enough to ensure that things looked the same to everyone. However, this is far from sufficient. A classic case is provided by Newton's famous laws of motion. Take the first law as an example. It tells us that bodies acted upon by no forces do not accelerate. They remain at rest or move at constant velocity. However, as Newton appreciated very clearly, this famous 'universal' law is not really universal. It will only be found to be true by a special class of observers in the Universe – those which are called 'inertial' observers. Inertial observers are those who are not accelerating or rotating relative to an imaginary cosmic background stage of space defined by the most distant stars.[24] These observers therefore violate the Copernican imperative. They see a Universe whose laws are especially simple. To see why this is so, imagine that you are located inside a spaceship out of whose windows you can observe the unchanging distant stars. Now suppose that the rocket boosters are fired so as to make the spaceship rotate. If you look out of the windows you will see the stars rotating (in the opposite sense) across the expanse of space. These stars will therefore appear to be accelerating[25] even though they are not being acted upon by any forces. Newton's law will not be seen to hold for this rotating, non-inertial, observer. By working a little harder the rotating observer can find the law that governs what he sees from his rotating vantage point but it is more complicated than the law seen by inertial observers. This undemocratic situation, that allowed some observers to see simpler laws of Nature than others, struck Einstein as a clear sign that there was something imperfect about the way Newton chose to express his laws of Nature. They could not be truly universal laws of Nature if they only held for special observers.

Einstein enunciated what he called the Principle of Covariance: that laws of Nature should be expressed in a form that will look the same for all observers, no matter where they are located and no matter how they are moving. When it came to implementing this Principle Einstein was very fortunate. During the latter part of the nineteenth century, pure mathematicians in Germany and Italy had been extremely busy developing a detailed understanding of all the possible geometries that could exist on curved surfaces. In doing that they had developed a mathematical language which automatically had the property that every equation possessed a form that remained the same if the co-ordinates describing it were changed in any way at all. This language was called the tensor calculus. Such changes of co-ordinates amount to asking what type of equation would be seen by someone moving in a different way. One of Einstein's oldest friends was a mathematician called Marcel Grossmann who was well informed about all these new mathematical developments. He showed Einstein this new mathematics of tensors and gradually Einstein realised that it was exactly what he needed to give a precise expression to his Principle of Covariance. So long as he expressed his laws of Nature as tensor equations they would automatically possess the same form for all observers.

This step by Einstein completes a dramatic movement in the physicist's conception of Nature that has been completed in the twentieth century. It is marked by a steady march away from there being any preferred view of the world, whether it be a human view, an Earth-based view, or a view based upon human standards. It has been brought about in stages. First, the Copernican revolution in astronomy gave rise to the view that our position in the Universe and the vantage point that we occupy in space and time is not specially privileged. Next, we have seen the creation of units of measurement and constants of Nature which are not reflections of human dimensions or the local astronomical motions of the Earth and the Sun. Instead, they are founded upon universal constants of Nature that transcend the human dimension. Last, we have seen how Einstein recognised that the laws of Nature themselves must be formulated in a way that ensures that any observer in the Universe, no matter where they are or how they are moving, should find the same laws to hold.

These steps have depersonalised physics and astronomy in the sense that they attempt to classify and understand the things in the Universe with reference only to principles that hold for any observer anywhere. If we have identified those constants and laws correctly then they provide us with the only basis we know upon which to begin a dialogue with extraterrestrial intelligences other than ourselves. They are the ultimate shared experience for everyone who inhabits our Universe.

chapter four

Further, Deeper, Fewer: The Quest for a Theory of Everything

'Physicists are trained to investigate a problem before arriving at a decision. Lawyers, advertisers and others are trained to do exactly the opposite: to seek data to confirm a determination that has already been made.'

Robert Crease[1]

NUMBERS YOU CAN COUNT ON

'An equation for me has no meaning unless it expresses a thought of God.'

Srinivasa Ramanujan[2]

Long ago, it became increasingly evident to our ancestors that Nature displayed both predictable and unpredictable events. The unpredictable aspects of things were dangerous and fearful. Perhaps they were punishments rained down by the gods to show their displeasure at human behaviour. They were also remarkable; as a result, ancient chronicles have a lot to say about plague, disaster and pestilence. Less newsworthy, but ultimately more significant, were the metronomic predictabilities of Nature. By noting and exploiting the periodic changes in the environment, crops could be grown, stocks laid in for the winter, and defences made against the incursions of wind and water. These regularities of Nature mirrored the regularities that structured stable societies and engendered a belief in law and order on a cosmic scale. Eventually, aided by the monotheistic faiths of many Western societies,[3] these ideas nurtured the idea that there exist things called 'laws of Nature' that hold at all times and in all places. These universal laws prescribe how things will behave not, like human laws, how they ought to behave.

We have come to appreciate that laws of change can always be replaced by the requirement that some other aspect of Nature does not change – called a conservation principle or an invariance of Nature. Energy is believed to be a prime example. It can be exchanged and reshuffled into different forms but, at the end of the day, when all the totting up is done the total must always be the same.

By the 1970s physicists had become so impressed by this correspondence between laws of Nature and unchanging patterns that they began to explore the catalogue of unchanging patterns to seek out candidates for the associated laws of change. Their search was extremely successful. The four basic forces of Nature – gravity, electricity and magnetism, radioactivity and nuclear interactions – were all described by working theories of this sort. Each of these four forces of Nature corresponds to a separate pattern which is preserved when anything happens in Nature: when a radioactive nucleus decays or a moving magnet in your bicycle dynamo produces an electric current.

All this was good news for physicists. By the mid 1970s they had separate theories of gravity, electromagnetism, the weak force (from which radioactivity derives) and the strong force (from which nuclear forces derive) which agreed with all observed events. The preservation of an unchanging pattern in each case required the respective force of Nature to exist and determined in detail how, and on what, it should act.

But still they were unhappy. Why should the world be governed by *four* different unchanging patterns? Even if your religious views include the notion of a Holy Quadrivium you might feel more instinctively drawn to regard *one* pattern and a single unified law of Nature as the most aesthetically, logically and physically appealing prospect. Any suggestion that the Universe might be a mixture of different laws that are unrelated in any respect smacks of a world that is a botched job. Of course, this is no proof that the Universe really is a harmonious single piece of legislation or a collection of occasionally conflicting principles.[4] Indeed, as the United States discovered about their Constitution after their presidential election in 2000, one might believe the former yet discover that the reality is closer to the latter. However, until there is real evidence to the contrary, scientists wisely assume that whatever is responsible for the patterns that we call the 'laws of Nature' is a good deal smarter than we are and will not have missed neat and beautiful patterns that are evident to us. Nor is this humble belief merely a piece of pious self-denial. It is based upon past experience. Time and again we have found that the laws of Nature are cleverer, more abstract, and less arbitrary than we had previously imagined.

This belief in the ultimate simplicity and unity behind the rules that constrain the Universe leads us to expect that there exists a single unchanging pattern behind the appearances. Under different conditions this single pattern will crystallise into superficially distinct patterns that show up as the four separate forces governing the world around us. It has gradually become clear how this patterning probably works.

We have learnt that the forces of Nature are not as distinct as they first appear. They seem to have very different strengths and to act upon different elementary particles. But this is an illusion created by our need to inhabit a place in the Universe where the temperature is rather low — low enough for atoms and molecules to exist. As the temperature rises and the elementary particles of matter collide with one another at higher and higher energies, the separate forces that govern our quiescent low-temperature world become more and more alike. The strong forces get weaker and the weak forces get stronger. New particles appear as higher temperatures are reached and they mediate interactions between the separate families of particles which, at low temperatures, appear to be isolated from one another. Gradually, as we reach those unimaginable conditions of the 'ultimate' temperature that Max Planck found defined by the four constants of Nature, G, k, c and h, we expect the distinctions to have been completely effaced and the forces of Nature will finally present a single united front.

COSMIC CUBISM

> 'There may be said to be two classes of people in the world: those who constantly divide the people of the world into two classes and those who do not.'

Robert Benchley[5]

The Soviet physicist George Gamow created a memorable fictional hero in a sequence of books that charted the exploits of Mr C.G.H. Tompkins, a bank clerk with an irrepressible interest in modern science[6] (see Figure 4.I).

Gamow's device for explaining the novel aspects of quantum physics and relativity was to create a fictional world where the effects were magnified enormously. In effect, this is done by changing the values of the constants of Nature. If the speed of light were to become 186 miles per hour rather than 186,000 miles per second[7] then the peculiar effects of motion upon the rate of passage of time and the measurement of distance would become features of everyday experience. You couldn't drive a car without being fully aware of them. Similarly, if Planck's constant were very much larger then the quantum wavelike aspects of matter would be constantly evident. When Mr Tompkins hits a billiard ball with his cue he finds that it takes many paths simultaneously, rather than the single definite path that they all

Figure 4.1 *The irrepressible Mr C.G.H. Tompkins, the eponymous hero of George Gamow's scientific fantasy,* Mr Tompkins in Wonderland.[8]

combine to produce in a world (like ours) where quantum effects are very small.[9]

Mr Tompkins's initials C.G.H. bear witness to the central importance of the constants of Nature characterising gravity (G), quantum reality (h), and light (c). We can use them to paint a simple picture of the correspondences between different laws of Nature. We need only to appreciate a simple principle. When G is set equal to zero we are turning off the force of gravity and ignoring it; when h is set equal to zero we are ignoring the quantum nature of the Universe, through which energies can only take on particular values, like steps on a ladder. The size of the steps between the rungs are fixed by h. If h were zero there would be no gaps and the energy of an atom could change by any value, no matter how small.[10] Third, when c is set equal to infinity (or, what is the same thing, $1/c$ equal to zero) then light signals move with infinite speed. This was the picture of the world in Newton's day, with gravity acting instantaneously between the Earth and the Sun.

Now we can create a three-dimensional picture of the possibilities by drawing a cube[11] whose axes measure the size of h, G and $1/c$, shown in Figure 4.2. Our cube has eight corners and each one represents a different physical theory. The simplest is at the origin of the graph where gravity is not included ($G = 0$), no quantisation is included ($h = 0$) and relativity is ignored ($1/c = 0$): this is Newtonian mechanics (NM). Moving vertically up the $1/c$ axis leaving $h = G = 0$, we encounter the theory of special relativity (SR). Moving horizontally along the h axis, leaving $1/c = G = 0$ we generalise Newton's mechanics to quantum mechanics (QM). Adding gravity by moving along the G axis we reach Newton's theory of gravity (NG). Moving upwards, leaving $h = 0$, we reach Einstein's general theory of relativity (GR), and it can also be reached by adding gravity to special relativity. Similarly, moving upwards from quantum mechanics by incorporating a finite value of $1/c$, we reach quantum field theory (QFT). Moving across the floor, so $1/c = 0$ still, we reach the quantum version of Newtonian gravity (NQG). Finally, the last unvisited corner of the cube is a theory that is relativistic, gravitational, and quantum (TOE). It is a generalisation of all other theories. It has yet to be found. So far physicists have identified a variety of so called 'string theories' which are limiting cases of a larger and deeper theory, dubbed M (for Mystery) theory. But the form of this deeper theory of which the known string theories are shadows cast in different directions is so far unknown.

The picture we have just created reveals a profound truth about the way in which progress occurs in science. Mature scientific progress is not a succession of revolutions which tear up old theories to make room for new ones. If that were true then the only thing about our current theories that we could be sure about is their incorrectness. Eventually, they will all be shown to be wrong. However, this cannot be the whole story. Those theories have been built upon millions of correct predictions. How can we take that into account in some way?

Newton's three-hundred-year-old theories of motion and gravity provide wonderfully accurate rules to understand and predict the way

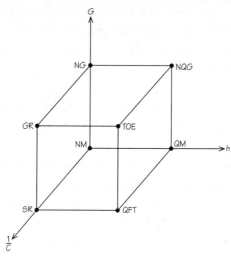

Figure 4.2 *How the structure of physical theories is controlled by the constants, G, c, and h. When G = 0, gravity is turned off; when h = 0, there is no quantisation of energy in Nature; when c equals infinity or 1/c = 0, there is no maximum speed for transmitting information and relativity is omitted. Plotting just representative non-zero or zero values of G, h, or 1/c we can identify the locations of increasingly general theories of physics.*

At the first level of generality we have:

 NM: Newton's mechanics (G = h = 1/c = 0).

At the second level we have:

 NG: Newton's theory of gravity (h = 1/c = 0, G ≠ 0)

 SR: Einstein's special relativity theory, which excludes gravity (h = G = 0, 1/c ≠ 0)

 QM: quantum mechanics (G = 1/c = 0, h ≠ 0).

At the third level we have:

 GR: Einstein's general theory of relativity, which adds gravity to special relativity (h = 0, G ≠ 0 and 1/c ≠ 0)

 QFT: Relativistic quantum mechanics, (G = 0, h ≠ 0, 1/c ≠ 0)

 NQG: Newtonian quantum gravity, (1/c = 0, h ≠ 0, G ≠ 0)

And ultimately, a yet to be found unified 'theory of everything':

 TOE: relativistic quantum gravity (1/c ≠ 0, h ≠ 0, G ≠ 0).

This diagram also illustrates how newer, larger theories contain their predecessors as limiting cases which can be recovered by taking an appropriate limit: 1/c → 0, h → 0, or G → 0.

things move at speeds far less than that at which light travels when gravity is very weak. In the first fifteen years of this century, Einstein found a deeper theory that could deal with fast motion and strong gravity when Newton's theory failed. But, crucially, Einstein's broader and deeper theory turns into Newton's when motions are slow and gravity is weak.

It was the same with the revolutionary quantum theories that were found in the first quarter of this century. They provided a more complete description than Newton of the way the world works when we probe the realm of the very small. Their predictions about the non-Newtonian microworld are stupendously accurate. But again, when they deal with large objects they become more and more like Newton's description of motion. This is how the core of truth within a past theory can remain as a limiting part of a new and better theory. Scientific revolutions don't seem to happen any more.

If we look at our cube of theories again we can see the inter-relationships between old and new theories. Take our case that quantum mechanics becomes Newtonian mechanics as h approaches zero. This limit corresponds to a situation in which the quantum wavelike aspects of particles become negligible. And this is why we can be completely confident that Newton's three-hundred-year-old theories of motion and gravity will be taught and used just as effectively in 1000 years time as they are today. Whatever the ultimate Theory of Everything is found to be, it will have a limiting form which describes motion at speeds far less than that of light in weak gravity fields where quantum wavelike features of mass are negligible. This form will be the one that Newton found.

NEW CONSTANTS INVOLVE NEW LABOUR

Einstein: 'You know, Henri, I once studied mathematics, but I gave it up for physics.'

Poincaré: 'Oh, really, Albert, why is that?'

Einstein: 'Because although I could tell the true statements from the false, I just couldn't tell which facts were the important ones.'

Poincaré: 'That is very interesting, Albert, because, I originally studied physics, but left the field for mathematics.'

Einstein: 'Really, why?'

Poincaré: 'Because I couldn't tell which of the important facts were true.'

Conversation between Albert Einstein and Henri Poincaré[12]

We have begun to see some of the ways in which the unveiling of new constants of Nature can help organise our understanding of the world. They are like beacons from which we can take our bearings. Real advances in our understanding of the physical world always seem to involve either:

(i) *Revelation:* The discovery of a new fundamental constant of Nature;

(ii) *Elevation:* The enhancement of the status of a known constant;

(iii) *Reduction:* The discovery that the value of one constant of Nature is determined by the numerical values of others;

(iv) *Elucidation:* The discovery that an observed phenomenon is governed by a new combination of constants;

(v) *Variation:* The discovery that a quantity believed to be a constant of Nature is not truly constant;

or

(vi) *Enumeration:* the calculation of the value of a constant of Nature from first principles, showing that its value is explained.

As an example of *revelation*, we recall how the introduction of the quantum theory by Planck, Einstein, Bohr, Heisenberg and others introduced us to the new fundamental constant, h, that bears Planck's name. It gave a finite numerical value to something that was previously assumed to be zero: the smallest energy change that can be observed in Nature.

Another more recent example is suggested by the development of a candidate for the title 'theory of everything', called superstring theory, in which the fundamental ingredients of the world are not point particles of mass but loops, or strings, of energy which possess a tension, rather like elastic bands. This string tension is the basic defining constant of the theory. Almost all other properties of the world follow from it (although they are yet to be worked out in most cases). This string tension may prove to be as fundamental as the Planck units of mass and energy.

As an example of *elevation*, we see how Einstein's development of the theory of special relativity gave a new universal status to the velocity of light in vacuum, c. Einstein showed that it provides the link between the concepts of mass (m) and energy (E) through his famous formula $E = mc^2$. Einstein did not discover that light moved with a finite speed. That had been observed long before and precise measurements of the speed of light had been made in the nineteenth century.

But Einstein's new theory of motion changed the status of the speed of light in vacuum forever. It became the ultimate speed limit. No information can spread faster. More fundamental still, it was the one velocity that all observers, no matter what their own motion, should always find to be the same. It was unique amongst all velocities.

The discovery of a *reduction* is something that usually comes later in the game than either *revelation* and *elevation*. We already need to know some candidate constants; then we need to develop a broader explanation that links together their domains of application. Often, the constants defining each of the areas that are made to overlap will be found to be linked. This is typically what happens whenever physicists manage to create a theory that 'unifies' two, previously distinct, forces of Nature. In 1967, a theory was proposed by Glashow, Weinberg and Salam that linked electromagnetism and the weak force of radioactivity. This theory was successfully tested by observation for the first time in 1983 and it links together the constants of Nature that label the strengths of the forces of electromagnetism and radioactivity. The links serve to reduce the number of independent constants that are believed to exist.

The discovery of an *elucidation* is slightly different to that of a *reduction*, but equally revealing. It occurs when a theory predicts that some observed quantity – a temperature or a mass, for example – is given by a new combination of constants. The combination tells us something about the inter-relatedness of different parts of science.

A good example is provided by Stephen Hawking's prediction, in 1974, that black holes are not entirely black. Thermodynamically, they are black bodies: perfect radiators of heat radiation. Prior to then it was believed that black holes were just cosmic cookie monsters, swallowing everything that came within their gravitational clutches. Once you fell inside a surface known as the event horizon, there was no return to the outside world.

Hawking succeeded in discovering what would happen if quantum processes were included in the story. Remarkably, black holes then turned out to be not quite black. The strong change in gravity near the event horizon could turn the gravitational energy of the black hole into particles which could be radiated away from the black hole, gradually sapping the mass of the hole, until it disappeared in a final explosion.[13] What is unusual about this evaporation process is that it is predicted to be governed by the simple everyday laws of thermodynamics that apply to all known hot bodies in equilibrium. Thus black holes turn out to be objects that are at once gravitational, relativistic, quantum mechanical and thermodynamical. The formula which gives the temperature of the radiation that a black hole of mass M radiates away into space by means of Hawking's evaporation process involves the constants G, h and c. But it also includes the thermodynamic constant of Boltzmann, k, which links energy to temperature. This is a spectacular elucidation of the interlinked structure of superficial disparate pieces of Nature.

The discovery of a *variation* is quite different to the previous four developments. It means that a quantity which we believed to be constant is discovered to be an imposter, masquerading as a true constant. It varies in space or in time. Generally, such a step will require the variation to be very small, or the quantity would not have been believed to have been constant in the first place. None of the fundamental constants of Nature have so far indubitably suffered this downgrading of their cosmic status. However, as we shall see later on, some are under suspicion as they have had their constancy probed to greater and greater levels of precision.

The prime suspect for tiny variations has always been the gravitational constant, G. Gravity is far and away the weakest force of Nature and the least closely probed by experiment. If you look up the known values of the major constants in the back of a physics textbook you will discover that G is specified to far fewer decimal places than c, h or e. In the mid 1960s it was thought for a time that Einstein's general theory of relativity disagreed with observations of the motion of the planet Mercury around the Sun. The first thing that was done to reconcile the two was to extend Einstein's theory by allowing G to change with time. Ultimately, the problem was traced to incorrect observations but, like a genie, once the varying-G theory was released it couldn't be shut up again.

Although G has withstood assaults on its constancy for longest, the most recent and detailed attacks have been launched against the constancy of α, the fine structure constant. They are so topical that we will be looking at them in more detail in Chapter 12. The fine structure constant is a linkage of the speed of light, Planck's constant and the electron charge. If it varies then we may choose to which of these we attribute the time variation.

All of these five touchstones of progress revolve around constants of Nature and they show the central role that constants play in our appraisal of progress. There is a sixth development on our list. We called it *enumeration*. This is the Holy Grail of fundamental physics and it means the numerical calculation of one of the constants of Nature. This has never been done. So far, the only way we can know their values is by measuring them.[14] This seems unsatisfactory. It allows the constants that appear in our theories to have a huge range of different possible values without overthrowing the theory. This is not the situation that Einstein imagined when we heard him speak in the last chapter. He thought that the true theory should only permit one choice for the constants that define it — the values we observe. Some people share his view today, but it has become increasingly apparent that not all the constants that define the world need be uniquely strait-jacketed in this way. It is likely that some are determined in a more liberal fashion by quantum randomness.

Many people hope that a complete theory would allow us to calculate the numerical values of some constants, like c, h and G, as accurately as we liked. This would also be a wonderful way of testing such a 'complete' theory. So far, this is just a dream. None of the constants that we believe to be truly fundamental has been calculated in this way from one of the theories in which it appears. Yet, such a calculation may not be too far away. Just a few years ago physicists were at an impasse with several possible string theories on offer, all seeming to be equally viable Theories of Everything. This was odd. Why did our Universe use just one of them? Then Ed Witten of Princeton University made a major discovery. He showed that all these superficially different string theories were not different at all. They were just different limiting situations of a single, bigger, deeper theory which we have yet to find. It is as if we are illuminating a strange object from many different angles, casting different shadows on a wall. From enough of these shadows it should be possible to reconstruct the illuminated object. This deep theory is the M theory introduced earlier in this chapter. Hidden within its mathematical defences is an explanation for the numerical values of the constants of Nature. So far, no one has been able to penetrate them and extract the information. We know a little about the structure of the M theory but the mathematics needed to elucidate it is formidable. Physicists are used to being able to take mathematics that mathematicians have already developed and use it like a tool to fashion physical theories. For the first time since Newton patterns have been encountered in Nature that require the development of new mathematics in order to further our understanding of them. Witten believes we have been lucky to stumble upon M theory about fifty years too early. Others might point to the warning that the most dangerous thing in science is the idea that arrives before its time.

Despite the lack of a fundamental theory with which to pursue a calculation of constants there has been no lack of numerological efforts to explain them. This is an activity that has a history, anthropology and sociology all of its own. Its fruits are rather unusual, and occasionally fantastic, as we are about to see.

NUMEROLOGY

'Here lies John Bun,
Who was killed by a gun,
His name was not Bun, but Wood,
But Wood would not rhyme with gun,
  but Bun would.'

Epitaph[15]

Lucky numbers, unlucky numbers, special numbers – lots of people think they can count on them. Here is a modern remnant of an ancient superstition. If we look back to about 550 BC we find Pythagoras and his Greek disciples carrying out some of the earliest studies of mathematics for its own sake. They were interested in everything about the Universe to which number could be attributed. This was a way of linking these disparate parts of the world together, making the planetary motions into a musical scale, and turning quantities into geometrical shapes. Unlike us they didn't think that numbers were just attributes of things. They thought that everything *was* number. Numbers had intrinsic meanings. They were not just relationships between things. From these religious beliefs there followed a quest to explore the numbers of things in all possible ways, looking for coincidental links between numbers in one area of life and another. Some numbers had good properties, others were bad. Some needed to be kept secret, others could be told to all.

To see how Pythagoras was led to this strong belief in numerology we should consider some of the games he liked to play with numbers. One of his favourites was the sequence of triangular numbers. Here we can see how a simple pattern of numbers can emerge rather naturally once pebbles or other counters are placed on the ground. If we place successive rows of one, two, three . . . dots below each other then we construct a progression of numbers which are 'triangular' in form (see Figure 4.3).

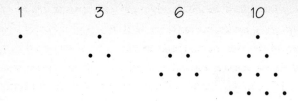

Figure 4.3 *Triangular numbers are created by laying down rows of dots with one more dot than in the row above.*[16]

Add them up row by row to form the progression of triangular numbers: 1, 1 + 2 = 3, 1 + 2 + 3 = 6, 1 + 2 + 3 + 4 = 10, and so on.[17]

This was particularly illuminating to the Pythagoreans because the Greeks denoted numbers by letters of their alphabet and this obscured the patterns in the sequence of numbers that are immediately evident to us. Pythagoras' pictorial representations of the triangles of numbers was fascinating. Indeed, we see a memory of it today when we refer to numbers as 'figures'. A figure still carries the dual meaning of a picture and a number. From this followed a picture of 1 as a point, 2 as a line joining two points, and 3 as a triangle, the first plane figure enclosing an area. The number 4 then symbolised the first solid figure, a pyramid composed of four triangular surfaces with four corner points.

In the same way, it was possible to speak of 'square' numbers, 4, 9, 16, 25 . . . which can be built up by dots laid out in square arrays. Alternatively, they noticed that they could be built up by adding together successive odd numbers, so for example,[18]

$$4 = 1 + 3$$
$$9 = 1 + 3 + 5$$
$$16 = 1 + 3 + 5 + 7$$
$$25 = 1 + 3 + 5 + 7 + 9$$
$$36 = 1 + 3 + 5 + 7 + 9 + 11$$

and so on.

These examples show how Pythagoras was drawn to make his first imaginative leap and regard numbers as things — geometrical objects. Next, he made an even more impressive discovery. He noticed that the tuning of Greek musical instruments depended upon simple numerical ratios, 1:2, 3:2, 4:3 and 8:9. These were the only musical intervals that the Greeks regarded as consonant and appealing to the ear. The impact of this discovery on Pythagoras' thinking was far-reaching. He thought that he had discovered that changes in human sense perceptions were dependent on mathematics. Moreover, the appearance of similar numbers in the description of musical intervals and the motion of the planets convinced the Pythagoreans that these superficially different phenomena are intimately linked.

What lies at the root of numerology is a belief that there is something intrinsically meaningful about the numbers themselves; that sevenness is a shared quality that links together all things that have a sevenfold quality, whether they are seven brides and seven brothers or the seven days of the week. It is a small step for some numbers, like 13, to be thought unlucky, or others, like 7, to be thought propitious. The Pythagoreans endowed certain numbers with special attributes, like goodness or justice. They became symbols in more ways than one. Here is a typical commentary:

> 'Because they assumed, as a defining property of justice, requital or equality, and found this to exist in numbers, therefore they said that justice was the first square number for in every kind the first instance of things having the same formula had in their opinion the best right to the name. This number some said was 4, as being the first square, divisible into equal parts and in every way equal, for it is twice 2. Others, however, said that it was 9, the first square of an odd number, namely 3 multiplied by itself.
>
> Opportunity, on the other hand, they said, was 7, because in nature the times of fulfilment with respect to

birth and maturity go in sevens. Take man for instance. He can be born after seven months, cuts his teeth after another seven, reaches puberty about the end of his second period of seven years, and grows a beard at the third.'[19]

Some numbers were especially revered because of their special properties. 'Perfect' numbers were so called because they have the remarkable property that they are equal to the sum of all the numbers that divide them exactly, apart from themselves. The first perfect number is $6 = 1 + 2 + 3$, the second is $28 = 14 + 7 + 4 + 2 + 1$. The next two are 496 and 8128 and were also known to the early Greeks. Even today only about 33 are known[20] and nobody knows if there are infinitely many of them, as there are prime numbers.[21]

Pythagoras was also much impressed by a succession of numbers which he called 'amicable'. Two numbers are called 'amicable' if the sum of the divisors of the first number is equal to the second number, and vice versa. In some sense they were judged to have the same 'parents' and the divine would look more favourably upon things that were numbered by pairs of these friendly numbers. For example 220 and 284 are amicable numbers.[22] You can divide 220 by 1, 2, 4, 5, 10, 11, 20, 22, 44, 55 and 110. Add them up and you get 284. You can divide 284 by 1, 2, 4, 71 and 142. Add them up and you get 220. Early Jewish scholars were very fond of using numerology to validate the texts of their scriptures or to extract some further hidden significance to quantities they contained.[23] This evolved into the most extreme forms of Kabbalism with its reverence for sevenfold occurrences. Here is a piece of numerological alternative medicine to cure malaria:

'take seven pickles from seven palm trees, seven chips from seven beams, seven nails from seven bridges, seven ashes from seven ovens, seven scoops of earth from seven door sockets, seven pieces of pitch from seven ships, seven hand-fuls of cumin, and seven hairs from the beard of an old dog, and tie them to the neck-hole of the shirt with a white twisted cord.'[24]

The most 'holy' Pythagorean numbers were the first four, 1, 2, 3 and 4, which formed the triangular number 10 (see Figure 4.4).

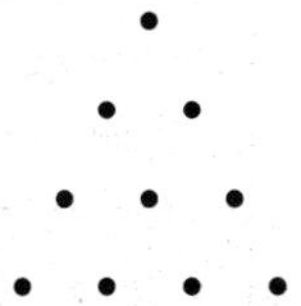

Figure 4.4 *The sacred tetraktys, the triangular representation of the number 10 as 1+2+3+4.*

This triangular representation of the number 10 was the symbol of the sacred tetraktys by which initiates into the Pythagorean order had to swear their oath of secrecy and allegiance. As part of their entry requirements they were sworn to secrecy for three years and as a result during the Renaissance the number of days in three years ($3 \times 365 = 1095$) was taken to be the number of silence. The tetraktys was nothing less than the master key to unlock our understanding of the whole of life and experience. Here is the account of one first-century commentator on the ten collections of four things that it was believed to symbolise:[25]

> 'Numbers: 1, 2, 3, 4.
> Magnitudes: point, line, surface, solid.
> Simple Bodies: fire, air, water, earth.
> Figures of simple bodies: pyramids, octahedron, icosahedron, cube.
> Living Things: seed, growth in length, in breadth, in thickness.
> Societies: man, village, city, nation.
> Faculties: reasons, knowledge, opinion, sensation.
> Seasons of the Year: spring, summer, autumn, winter.
> Ages: infancy, youth, manhood, age.
> Parts of the human being: body and the three parts of soul.'

These curious ideas were extraordinarily persistent. In every age, in every place, there were writers and thinkers who were fascinated by the meaning of numbers. They treated equations and formulae as if they were secret codes that encrypted the true meaning of the Universe. Nor has this view become extinct today. Although we use mathematics to establish relationships between things, there is still a population of amateur investigators who are seeking a special 'formula' that will tell us something about the ultimate nature of the physical world. And what better thing for this formula to tell us than those numbers that lie at the heart of physical reality: the values of the constants of Nature. Numerology has turned its attention to the physical constants of Nature in an attempt to explain their values in a concatenation of πs, square roots and common numbers.

These efforts feed on coincidences. Some of the most impressive have no discernible significance at all. For instance,[26] it was once noted that

$$\exp\{\pi(\sqrt{67})/3\} \approx \text{the number of feet in one mile}$$

to an accuracy of 1 part in 300 million! Or how about the claim that the number $\exp\{\pi(\sqrt{163})$ is an integer, first made by Charles Hermite in 1859. It is known to be extraordinarily close, to be

$$262,537,412,640,768,743.999,999,999,999,25 \ldots$$

This formed part of an April Fool's hoax by Martin Gardner who claimed that it *was* an integer and that the Indian mathematician Ramanujan had predicted it.[27] As a result it has become known as 'Ramanujan's Constant'.

But there are an awful lot of numbers and even more possible permutations of them. Coincidences seem more striking because we don't think about how many unimpressive 'non-coincidences' we encounter in between finding them. When analysed from a statistical perspective it turns out that coincidences like this are not that unusual. Remember when Uri Geller used to appear on television and announce that he was going to stop the clocks in your home. There were millions of viewers and we expect quite a lot of wind-up clocks will be stopping as he speaks. Those households where a clock stops are terribly impressed. The rest just think they can't have been psychically tuned in enough. After all, he did manage to stop a lot of other clocks.

My favourite numerical coincidence is one that my literary friend Stephen Medcalf told me about as an example that would defy any attempt to evaluate the mathematical likelihood of it arising by chance. I think it was noticed by an Eton schoolboy about seventy years ago. First, a little background which I am not able to judge. There is a tradition, or legend, that William Shakespeare had a hand in producing the English renderings of some of the Psalms in the Authorised King James Version of the English Bible.[28] It was suggested that his hand is detectable in the Psalm 46, written in the year when Shakespeare was 46. For, as the schoolboy noticed, the 46th word from the beginning of this psalm[29] is 'shake'. The 46th from the end is 'spear'. Coincidence or hidden signature?

All sorts of numerical coincidences which incorporate the values of some of the constants of Nature can be found in the literature of science and vastly more in the in-trays of physicists where they have arrived from well-meaning correspondents. Here are a few of the proposed formulae (none are taken seriously) for the fine structure constant. Compare them with the best experimental value:

experimental: $1/\alpha = 137.035989561 \ldots$

First, there have been attempts to 'prove' that $1/\alpha$ equals the following expressions using a speculative extension of known physics:

Lewis and Adams[30] $1/\alpha = 8\pi(8\pi^5/15)^{1/3} = 137.348$
Eddington[31] $1/\alpha = (16^2 - 16)/2 + 16 + 1 = 137$

Wyler[32] $1/\alpha = (8\pi^4/9)(2^4 5!/\pi^5)^{1/4} = 137.036082$
Aspden and Eagles[33] $1/\alpha = 108\pi(8/1843)^{1/6} = 137.035915$

Of course, if M theory eventually comes up with a determination of the value of $1/\alpha$ it might well look rather like one of these speculative formulae. However, it would supply a large and consistent theoretical edifice from which the prediction would follow. It would also need to make some predictions of things that we have not yet measured, for example the next few decimal places of $1/\alpha$, that future experimenters could search out and check.

All these pieces of numerical gymnastics are impressively close to the experimental value (they were even closer in the past when they were first proposed) but the prize for persistent ingenuity must go to Gary Adamson,[34] whose rogues' gallery of 137-ology is shown in Figure 4.5.

These examples at least have the virtue of emerging from some attempt at a theory of electromagnetism and particles. But there are also 'pure' numerologists who look for any combinations of powers of small numbers and weighty mathematical constants like π which get close to the required 137.035989561. . . . Here are a few examples of this sort:

Robertson[35] $1/\alpha = 2^{-19/4}3^{10/3}5^{17/4}\pi^{-2} = 137.03594$
Burger[36] $1/\alpha = (137^2 + \pi^2)^{1/2} = 137.0360157$

Even the great theoretical physicist, Werner Heisenberg, couldn't resist a tongue-in-cheek suspicion that[37]

'As to the numerical value I suppose $1/\alpha = 2^4 3^3/\pi$, but that is of course in play.'

This is more than enough of numerology. After a while it starts to become addictive. It is easy to see why it has been so universally

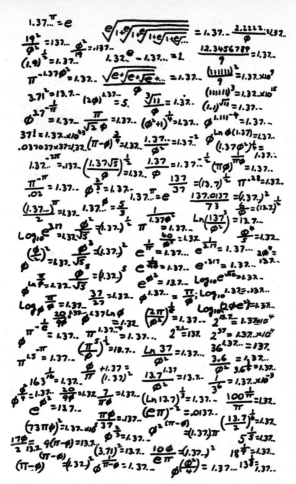

Figure 4.5 *Some numerological flights of fancy involving the number 137, compiled by Gary Adamson.* $\phi = 1.61803 \ldots$ *is the Golden Ratio.*

enduring in its fascination. Our purpose in revealing some of its examples is not without a serious object. One of the formulae we listed above bears the name of Arthur Eddington, one of the twentieth century's greatest astrophysicists. In the next chapter we will look at Eddington more closely. He is a remarkable combination of the profound and the fantastic. More than any modern figure, he is responsible for setting in motion the never-ending attempts to explain constants of Nature by feats of pure numerology. He also noticed a new and dramatic feature of the constants of Nature.

chapter five

Eddington's Unfinished Symphony

'I have had a most rare vision, I have had a dream, – past the wit of man to say what dream it was: man is but an ass if he go about to expound this dream . . . It shall be called Bottom's dream, because it hath no bottom.'

A. S. Eddington[1]

COUNTING TO 15,747,724,136,275,002,577, 605,653,961,181,555,468,044,717,914,527,116,709, 366,231,425,076,185,631,031,296

'Conservatism is suspicious of thinking, because thinking on the whole leads to wrong conclusions, unless you think very, very hard.'

Roger Scruton[2]

'Any coincidence is always worth noticing,' Miss Marple told us; after all, 'you can throw it away later if it is *only* a coincidence.' One of the most striking features about the study of the astronomical universe during the twentieth century has been the role played by coincidence: its existence, its neglect and its recognition. As physicists started to appreciate the role of constants in the quantum realm and to exploit

Einstein's new theory of gravity to describe the Universe as a whole the time was ripe for someone to try to marry the two together.

Enter one Arthur Stanley Eddington: a remarkable scientist who had been the first to discover how the stars were powered by nuclear reactions. He also made important contributions to our understanding of the galaxy, wrote the first systematic exposition of Einstein's theory of general relativity and was responsible for one of the decisive experimental tests of Einstein's theory. He led one of the two expeditions to measure the tiny bending of light by the Sun's gravity, only measurable during a complete eclipse of the Sun. Einstein's theory predicted that the gravity field of the Sun should deflect passing starlight en route to Earth by about 1.75 seconds of arc as it passed by the Sun's surface. By taking a picture of a distant star field when the Sun's disc was covered by the Earth's shadow and again when the Sun was on the other side of the sky, any tiny shift in the apparent positions of the stars could be detected and the light-bending prediction tested. Eddington's team made a successful measurement in Principe despite poor weather conditions. His confirmation of Einstein's prediction was what launched Einstein into the public eye as the greatest scientist of the age. In Figure 5.1 they are seen together on the occasion of Einstein's visit to Cambridge, talking together in Eddington's garden at the University Observatories.

Eddington made a visit to Cal Tech in Pasadena in 1924 and found that his explanations of relativity, together with his experimental confirmation of its light-bending predictions, had coupled his name to Einstein's. An extremely modest and retiring character, he was delighted to find that the astronomers had not only organised a dinner in honour of his visit but that one of the physicists with whom he played golf had written a marvellous parody of *The Walrus and the Carpenter* to celebrate their mutual appreciation of relativity, golf and Lewis Carroll — who couldn't have done it better himself.

Figure 5.1 *Albert Einstein and Arthur Eddington together in Eddington's garden in 1930, photographed by Eddington's sister.*[3]

The Einstein and the Eddington

The Einstein and the Eddington
Were counting up their score
The Einstein's card showed ninety-eight
And Eddington's was more,
And both lay bunkered in the trap
And both stood up and swore.

I hate to see, the Einstein said
Such quantities of sand;

Just why they placed a bunker here
I cannot understand,
If one could smooth this landscape out
I think it would be grand.

The time has come, said Eddington,
To talk of many things;
Of cubes and clocks and meter-sticks,
And why a pendulum swings,
And how far space is out of plumb,
And whether time has wings.

And space it has dimensions four,
Instead of only three.
The square on the hypotenuse
Ain't what it used to be.
It grieves me sore, the things you've done
To plane geometry.

You hold that time is badly warped,
That even light is bent;
I think I get the idea there,
If this is what you meant:
The mail the postman brings today,
Tomorrow will be sent.

The shortest line, Einstein replied,
Is not the one that's straight;
It curves around upon itself,
Much like a figure of eight,
And if you go too rapidly
You will arrive too late.

But Easter day is Christmas time
And far away is near,
And two and two is more than four
And over here is near.
You may be right, said Eddington,
It seems a trifle queer.'

W.H. Williams

Eddington was a complex personality[4] with simple tastes. He was a serious-minded Quaker and a pacifist. His non-combatant position during the First World War led to him being chosen to lead the Principe eclipse expedition. As his career developed he came increasingly into the public eye through a series of impressively lucid popular science books that expounded the developing scientific view of the world, together with his own philosophy of science. His writings about the beginning and the end of the world inspired many writers to introduce scientific ideas into their plots, while theologians and philosophers were variously challenged and informed about the inevitability of the impending Heat Death of the Universe. Dorothy Sayers's Peter Wimsey story *Have His Carcase*[5] makes amusing use of the second law of thermodynamics and the steady increase of disorder it requires to reassure a muddled witness that such evidence is attuned to the thermodynamic way of the world. The role of the 'Second Law' in the evolution of the Universe was an important theme in Eddington's popular writings at the time. Here is what Sayers invents. A witness is worried that her evidence is so confused that no one will believe her . . .

'"But you do believe me, don't you?"

"We believe in you, Miss Kohn," said Wimsey, solemnly, "as devoutly as in the second law of thermo-dynamics."

"What are you getting at?" said Mr Simons, suspiciously.

"The second law of thermo-dynamics," explained Wimsey, helpfully, "which holds the universe in its path, and without which it would run backwards like a cinema film wound the wrong way."

"No, would it?" exclaimed Miss Kohn, rather pleased.

"Altars may reel," said Wimsey, "Mr Thomas may abandon his dress-suit and Mr Snowden renounce Free Trade, but the second law of thermo-dynamics will endure while memory holds her seat in this detracted globe, by which Hamlet meant his head but which I, with a wider intellectual range, apply to the planet which we have the rapture of inhabiting. Inspector Umpelty appears shocked, but I assure you that I know no more impressive way of affirming my entire belief in your absolute integrity." He grinned, "What I like about your evidence, Miss Kohn, is that it adds the final touch of utter and impenetrable obscurity to the problem which the Inspector and I have undertaken to solve. It reduces it to the complete quintessence of incomprehensible nonsense. Therefore, by the second law of thermo-dynamics, which lays down that we are hourly and momently progressing to a state of more and more randomness we receive positive assurance that we are moving happily and securely in the right direction."'

Whereas Eddington was a shy man with little talent for public speaking, he could write beautifully and his metaphors and analogies can still be found, used again and again by astronomers seeking graphic explanations for complicated ideas. He never married and lived at the Observatories in Cambridge where his sister kept house for him and their elderly mother. His interests were conservative but not totally predictable; he liked detective stories and football (the real sort that is played with a spherical ball), and would enjoy joining the crowds of London workers at Highbury to watch Arsenal, the leading team of his day.[6] He was an indifferent golfer and tennis player but was more serious about his cycling expeditions. His cycling record was coded in a single 'Eddington number', E, where E was the number of days in which he had cycled more than E miles. As E starts to get large, a very considerable effort is needed to increase it by even 1. At the time of his death Eddington's E number had grown to 87.

Eddington held the Plumian Professorship of Experimental Philosophy at Cambridge University. This antique title was by tradition that of the senior astronomer in the university. For part of his time in this post Eddington was a contemporary of Paul Dirac, the Lucasian professor of mathematics and youthful winner of the Nobel prize for physics. Dirac was one of the great physicists of the twentieth century, who predicted the existence of antimatter, developed the most transparent presentation of quantum mechanics, uncovered how the electron behaved, and much else besides. His work was of the most fundamental mathematical sort and performed entirely alone. He had no collaborators, only nominal research students, and no research group.

It was within this local climate of searches for new laws of Nature and the behaviour of its most elementary particles that Eddington began a programme of work that attracted the entire spectrum of responses, from awed admiration to open ridicule, from his peers. He called it his 'Fundamental Theory' and it was a search for the most basic possible physical theory, one that could explain the numerical values of the constants of Nature.

Eddington believed that pure thought could succeed in arriving at a complete description of the physical world. This was an even more ambitious idea in his day than it is today. Here is a brief expression of his credo:

> 'My conclusion is that not only the laws of nature but the constants of nature can be deduced from epistemological considerations, so that we can have a priori knowledge of them.'[7]

This is the ultimate manifesto of the theorist. Eddington believed that by pure thought it should be possible to deduce all the laws and constants of Nature and predict the existence of things like stars and galaxies in the Universe. The image he liked best was that of an astronomer on a cloud-covered planet who deduced the existence of the unseen stars above. Of course, experiments and observations made the task far easier but Eddington believed that was *all* they did. Without them, his goal would be harder to reach but not unattainable.

Eddington's programme was never completed. His book was unfinished[8] at the time of his death in 1944, but in the years before he had published a number of articles and devoted sections of his popular books to announcing great advances in his programme to understand the values of the constants of Nature. He focused his attention upon a small number of constants of Nature, raising their profile and challenging scientists to explain them, before embarking upon a complex chain of idiosyncratic mathematical reasoning that was designed to explain their values – exactly.

FUNDAMENTALISM

'In ancient days two aviators procured to themselves wings. Daedalus flew safely through the middle air across the sea, and was duly honoured on his landing. Young Icarus soared upwards towards the Sun until the wax melted which bound his wings, and his flight ended in fiasco. In weighing their achievements perhaps there is something to be said for Icarus. The classic authorities tell us that he was only "doing a stunt", but I like to think of him as the man who brought to light a constructional defect in the flying machines of his day.'

Arthur Eddington[9]

Eddington first took up his quest to explain the pure numbers that define our Universe in 1921, in the pages of his famous textbook on general relativity. He proposes that the characteristics of elementary particles of Nature like the electron should derive locally from the structure of the space and time in which they reside,[10] so that there must exist an unknown equation that expresses this relationship in the form:

'radius of electron in any direction = numerical constant ×
radius of curvature of space-time in that direction'

Amongst the numbers that Eddington regarded as of prime importance was the so called 'Eddington number' which is equal to the number of protons in the visible Universe.[11] Eddington calculated[12] this number to enormous precision (by hand) on a transatlantic boat crossing, concluding with the memorable statement that,

> 'I believe that there are 15,747,724,136,275,002,577,605, 653,961,181,555,468,044,717,914,527,116,709,366,231, 425,076,185,631,031,296 protons in the universe and the same number of electrons.'

This huge number, usually denoted by N_{Edd}, is approximately equal to 10^{80}. What attracted Eddington's attention to it was the fact that it must be a whole number and so it could in principle be calculated *exactly*.

During the 1920s when Eddington began his quest to explain the constants of Nature there was no good understanding of the weak and strong forces of Nature and the only dimensional constants of physics that were known and interpreted with confidence were those defining gravity and electromagnetic forces. Eddington arranged them into three pure dimensionless numbers. Using the experimental values of the day, he took the ratio of the masses of the proton and electron

$$m_{pr}/m_e \approx 1840,$$

the inverse of the fine structure constant

$$2\pi hc/e^2 \approx 137$$

and the ratio of the gravitational force to the electromagnetic force
between an electron and a proton,

$$e^2/Gm_{pr}m_e \approx 10^{40}$$

To these he added his cosmological number, $N_{Edd} \approx 10^{80}$. These four
numbers he called 'the ultimate constants'[13] and explaining their values
presented theoretical science with its greatest challenge:

> 'Are these four constants irreducible, or will a further unifi-
> cation of physics show that some or all of them can be
> dispensed with? Could they have been different from what
> they actually are? . . . the question arises whether the above
> ratios can be assigned arbitrarily or whether they are
> inevitable. In the former case we can only learn their values
> by measurement; in the latter case it is possible to find
> them by theory . . . I think the opinion now widely prevails
> that the [above four] constants . . . are not arbitrary but
> will ultimately be found to have a theoretical explanation;
> though I have also heard the contrary view expressed.'[14]

Speculating further, Eddington thought that the number of unexplained
constants was a helpful gauge of the gap to be closed before a truly
unified theory of all Nature's forces was unveiled. As to whether this
ultimate theory would contain one constant, or none at all, we would
have to wait and see:

> 'Our present recognition of four constants instead of one
> merely indicates the amount of unification of theory which

still remains to be accomplished. It may be that the one remaining constant is not arbitrary but of that I have no knowledge.'[15]

Eddington hoped that he could create a theory that would weave together the macroscopic world of astronomy and cosmology with the sub-atomic world of protons and electrons. His cosmic numbers were unusual in many respects. First, of course, no one had any idea why they took the particular numerical values that they did. Second, they span a huge size range. The proton-electron mass ratio and the fine structure constant are not too far away from pure numbers close to one and they might plausibly turn up as small products of numbers like 2, 3 or π in mathematical formulae. This is what Eddington was hoping for. But the other two numbers he selected are completely different. They are huge. The appearance of a number like 10^{40} in a formula in physics needs a very special explanation, or at least a reason that is very different from those that we are used to finding for things in science. Worse still, the vastly bigger number, $N_{Edd} \approx 10^{80}$, not only suffers an even bigger credibility problem, but it is tantalisingly close to being the square of the first large number. Surely this can't be a pure coincidence?! Eddington felt that if there was to be one number remaining as the defining quantity behind our Universe then that 'arbitrary constant' lay at the root of these huge numbers.[16] Of N_{Edd}, the largest and most mysterious number, he wrote:

'Regarded as the number of particles in the universe, it has generally been looked upon as a special fact [rather than as a necessary truth]. A universe, it is held, could be made with any number of particles; and so far as physics is concerned we must just accept the number allotted to our universe as an accident or as a whim of the Creator. But the epistemological investigation changes our idea of its nature. A universe cannot be made with a different number

of elementary particles consistently with the scheme of defi-
nitions by which the "number of particles" is assigned to
a system in wave mechanics. We must therefore no longer
look on it as a special fact about the universe, but as a
parameter occurring in the laws of nature and, as such,
part of the laws of nature.'[17]

We shall have a lot more to say about the 'large' numbers because
they played an influential role in the development of many cosmological theories. Eddington didn't have a theory that could explain them,
but he worked very hard on theories which might explain the smaller
numbers that lie close to 137 and 1840. These numbers control almost
all the gross features of atoms and atomic structures.

How did Eddington try to explain these numbers? One consistent
path of attack for all his calculations was to justify his special equation

$$10m^2 - 136m + 1 = 0$$

This equation is of a sort that is encountered first in school when you
are about 15 years old. There are two possible solutions of the equation[18] and they are in the ratio of 1847.6 to 1. This was close enough
to the proton-electron mass ratio (which even in Eddington's day was
nearer to 1836) to inspire Eddington to find any and every justification
for his equation and to seek out small adjustments that might explain
the small 'discrepancies'. The form of the equation itself he believed to
be dictated by the number of possible combinations and permutations
of numbers and directions that characterised our four dimensions of
space and time. The quantities 1, 10 and 136 that appear in the equation are 'derived' from the fact that there are $3^2 + 1^2 = 10$ simple quantities available to describe space and time and then $10^2 + 6^2 = 136$ at
the next most complicated level. At first Eddington had seized upon the
136 as likely to be the explanation for the value of the reciprocal of the
fine structure constant. But gradually he became persuaded that it was

necessary to multiply this number by $137/136$ (to get 137!) because of a mysterious argument about the need to take into account that the effective electric charges of two particles have an indistinguishable aspect to them. He claims that 'there is nothing mystical about indistinguishability'[19] but unfortunately almost everyone else thought there was.

This sequence of deductions created quite a stir of interest and criticism in scientific circles, both for the suspicious 'fudge' factor that shifted 136 to the more plausible 137 after the fact, and the stubborn experimental fact that unfortunately the fine structure constant did not appear to be an exact whole number at all. Eddington even wrote an article for one of the London newspapers explaining the problems of his esoteric deductions. Many other scientists were completely mystified and some, like Vladimir Fock, were moved to poetry about it all:[20]

> 'Though we may weigh it as we will,
> Exhausted and delirious
> *One-hundred-and-thirty-seven* still
> Remains for us mysterious.
> But Eddington, *he*, sees it clear,
> Denouncing those who tend to jeer;
> It is the number of (says he)
> The world's dimensions. *Can it?*! *be?*!'

Eddington's approach to the large numbers was not quite so obscure. It was certainly speculative but at least his colleagues could understand him. He was hoping that the masses of particles like the electron might derive in some way from the statistical fluctuations of all the masses in the universe. The magnitude of statistical fluctuations in collections of N particles is typically given by the square root of N and hence one might be persuaded that the ratio of the electric force to the gravitational force between two protons was a statistical fluctuation determined in magnitude by the square root of $N_{Edd} \approx 10^{80}$, which is very nearly 10^{40}.

THEATRICAL PHYSICS

'Analogies prove nothing, that is quite true, but they can make one feel more at home.'

Sigmund Freud

Eddington's methodology was mercilessly spoofed by other sceptical physicists of the day. Here is a charming example that Beck, Bethe and Riezler managed to fool the serious-minded editor of *Naturwissenschaften* into publishing[21] in German in 1931:

> ### Remark on the Quantum Theory of Zero Temperature
>
> We consider a hexagonal crystal lattice. The absolute zero of this is characterised by the condition that all degrees of freedom of the system freeze, that is all internal movements of the lattice cease. An exception to this is, of course, the motion of the electron in its Bohr orbit. According to Eddington each electron possesses $1/\alpha$ degrees of freedom, where α is the Sommerfeld fine structure constant. Besides electrons, our crystal contains only protons, and the number of degrees of freedom for them is the same since, according to Dirac, a proton can be regarded as a hole in the electron gas. Thus, since one degree of freedom remains because of the orbital motion, in order to attain absolute zero we must remove from a substance $2/\alpha - 1$ degrees of freedom per neutron (= 1 electron + 1 proton; since our crystal has to be electrically neutral overall). We obtain therefore for the zero temperature T_0
>
> $$T_0 = - (2/\alpha - 1)\text{Degrees}.$$
>
> Setting $T_0 = -273°$ we obtain for $1/\alpha$ the value 137, which, within limits of error, agrees completely with the value

obtained in an independent way. One can easily convince oneself that our result is independent of the special choice of crystal structure.

Cambridge, 10 December 1930

G Beck, H Bethe, W Riezler

Indeed, so convincing did this nonsense appear to some readers that Riezler was asked to present an exposition of the work in Munich at Sommerfeld's[22] weekly physics seminar. Eddington, however, was not amused, and nor was Herr Berliner, the editor of the journal, when he discovered that he had been made to look an ass. The serious-minded Herr Berliner immediately published an 'erratum' on 6 March which pointed out that

> 'The Note by G. Beck, H. Bethe and W. Riezler, published in the 9 January issue of this journal, was not meant to be taken seriously. It was intended to characterise a certain class of papers in theoretical physics of recent years which are purely speculative and based on spurious numerical arguments. In a letter received by the editors from these gentlemen they express regret that the formulation they gave this idea was suited to produce misunderstandings.'

But the mischievous George Gamow was not one soon to tire of a good joke and shortly afterwards he, Rosenfeld, and Pauli wrote separate letters from different European addresses protesting to the editor that the journal had now published another one of these disgraceful spoof articles and pointed the finger at another semi-numerological article, '*Origin of Cosmic Penetrating Radiation*', by some poor unsuspecting author,[23] demanding that the editor obtain its immediate withdrawal from the author in order to maintain the standards of the journal.

Here is another lampoon from Max Born's 1944 lectures on *Experiment and Theory in Physics:*[24]

> 'Eddington connects the dimensionless physical constants with the number n of the dimensions of his E spaces and his theory leads to the function $f(n) = n^2(n^2 + 1)/2$ which, for consecutive even numbers $n = 2, 4, 6, \ldots$ assumes the values $10, 136, 666 \ldots$ Apocalyptic numbers, indeed. It has been proposed that certain well-known lines of St. John's Revelation ought to be written in this way: "And I saw a beast coming up out of the sea having $f(2)$ horns and his number is $f(6) \ldots$" but whether the figure x in "$\ldots$ and there was given to him authority to continue x months $\ldots$" is to be interpreted as $1 \times f(3) - 3 \times f(1)$ or as $[f(4)-f(2)]/3$ can be disputed.'

An aside which illustrates the difficulty many had in reconciling Eddington's work on fundamental constants with his monumental contributions to general relativity and astrophysics can be found in a story told by Sam Goudsmit[25] about himself and the Dutch physicist, Kramers:

> 'the great Arthur Eddington gave a lecture about his alleged derivation of the fine structure constant from fundamental theory. Goudsmit and Kramers were both in the audience. Goudsmit understood little but recognised it as farfetched nonsense. Kramers understood a great deal and recognised it as total nonsense. After the discussion, Goudsmit went to his friend and mentor Kramers and asked him. "Do all physicists go off on crazy tangents when they grow old? I am afraid." Kramers answered, "No Sam, you don't have to be scared. A genius like Eddington may perhaps go nuts but a fellow like you just gets dumber and dumber."'

The most interesting thing about Eddington's attempts to explain the constants of Nature by algebraic and numerical gymnastics is their enduring effects on the readers of his popular books. He liked to tell his general readers about his new 'calculations' of the constants of Nature and the overwhelming impression he conveyed was that it might be possible to unlock some of the most deeply hidden secrets of the Universe by a little bit of inspired guesswork and numerology. If you noticed that some equations had solutions that lay close to numbers like 137 and 1840 then you were in business as a rival to Einstein. No need to observe the world and no need to make any predictions about things that had not yet been seen; the name of the game was linking numbers.

I believe that Eddington's work, and his widespread popularisation of it in books that sold in huge quantities and continued to be read for more than 60 years after they were first published, inspired a generation of amateurs who dreamed of finding *the* numerological explanation for the constants of Nature. Every week I receive letters that contain calculations of a sort that owe much to Eddington's style and approach to Nature. They are characterised by very detailed numerical calculations, a confinement of interest to a small subset of the constants of Nature, and no desire to predict anything new.

In order to evaluate the significance of relationships like these, or the ones that we saw proposed in the last chapter to explain the numerical value of the fine structure constant, we need to ask a simple question. What is the chance that seemingly impressive formulae arise purely by chance? If we pick a few suggestive numbers like 2, 3 or π and multiply them together a few times, how likely are we to get a number that agrees closely with some constant of Nature? Unfortunately for the numerologists, the answer is that these formulae are not very surprising at all.[26] It is easy to be impressed by numerical formulae because it is hard for us to realise how much deliberate searching has gone into the formula on display and how many ways there are to get a close agreement. For example, with a little work we can find the sort of formula that any modern-day Pythagorean would be proud of:[27]

$$666 + 6 + 6 + 6 = (6 - 6/6)^{(6 + 6 + 6)/6} + 6^{(6 + 6 + 6)/6}$$
$$+ (6 + 6/6)^{(6 + 6 + 6)/6}$$

But it should not be endowed with any apocalyptic significance.

Eddington's detailed attempts to explain the values of the constants of Nature have not led to a successful pathway of explanation; but they opened up new vistas and possibilities. They raised the sights of physicists and created a new frontier to strive for. His perennial rival James Jeans captured the significance of this unfulfilled quest perfectly when he wrote in 1947 of Eddington's unsuccessful search for a fundamental theory:

> 'Few if any of Eddington's colleagues accepted his views in their entirety; indeed few if any claimed to understand them. But his general train of thought does not seem unreasonable in itself, and it seems likely that some such vast synthesis may in time explain the nature of the world we live in, even though the time may not be yet.'[28]

Eddington's attempts to produce a unified explanation for the constants of Nature attracted few adherents. The great physicists of his day, like Dirac, Einstein, Bohr and Born, found it useless and politely confessed that they couldn't understand it. Eddington was frustrated by this reception and could not understand why others did not see things as he did, complaining in 1944 to his friend Herbert Dingle that[29]

> 'I am continually trying to find out why people find the procedure obscure. But I would point out that even Einstein was considered obscure, and hundreds of people have thought it necessary to explain him. I cannot seriously

believe that I ever attain the obscurity that Dirac does. But in the case of Einstein and Dirac people have thought it worthwhile to penetrate the obscurity. I believe they will understand me all right when they realize they have got to do so – and when it becomes the fashion "to explain" Eddington.'

That day never came.

chapter six

The Mystery of the Very Large Numbers

'History is the science of things which are not repeated.'

Paul Valéry[1]

SPOOKY NUMBERS

'Although we talk so much about coincidence we do not really believe in it. In our heart of hearts we think better of the universe, we are secretly convinced that it is not a slipshod, haphazard affair, that everything in it has meaning.'

J.B. Priestley

The greatest mystery surrounding the values of the constants of Nature is without doubt the ubiquity of certain huge numbers that seem to appear in a variety of apparently quite unrelated considerations. Eddington's number is a notable example. The total number of protons that lie within the encompass of the observable Universe[2] is close to the number

$$10^{80}$$

Then, if we ask for the ratio of the strengths of electromagnetic and gravitational forces between two protons, the answer does not depend on their separation,[3] but is equal to approximately

$$10^{40}$$

This is slightly sinister. It is peculiar enough for pure numbers involving the constants of Nature to turn out very different from numbers within a factor of a hundred or so of 1, but 10^{40} and its square, 10^{80}, is bizarre! Nor does it end there. If we were to follow Max Planck and compute an estimate for the 'action'[4] of the observable universe in units of the fundamental Planck units of action,[5] then we get

$$10^{120}$$

We have already seen that Eddington was inclined to relate the number of particles in the observable universe to some quantity involving the cosmological constant. This quantity has had a fairly quiet history since that time, occasionally re-emerging when theoretical cosmologists need to find a way of accommodating awkward new observations. Recently this scenario has been played out again. New observations of unprecedented reach and accuracy, made possible by the Hubble space telescope working in co-operation with sensitive ground-based telescopes, has detected supernovae in far distant galaxies. Their characteristic brightening and fading pattern allows their distance to be deduced from their apparent brightnesses. Remarkably, they turn out to be receding from us far more quickly than anyone expected. The expansion of the universe has turned from a state of deceleration into one of acceleration. These observations imply the existence of a positive cosmological constant. If we express its numerical value as a pure dimensionless number by measuring in units of the square of the Planck length, then we get a number very close to

$$10^{-120}$$

No smaller number has ever been encountered in a real physical investigation.

What are we to make of all these large numbers? Is there something cosmically significant about 10^{40} and its squares and cubes?

A BOLD HYPOTHESIS

'Look what happens to people when they get married!'

George Gamow[6]

The appearance of some of these large numbers had been a source of amazement ever since they were first noticed by Hermann Weyl in 1919. Eddington had tried to build a theory that made their appearance understandable. But he failed to convince a significant body of cosmologists that he was on the right track. Yet Eddington succeeded in persuading people that there was something that needed explaining. Completely unexpectedly, it was one of his famous neighbours in Cambridge who wrote the short letter to the journal *Nature* which succeeded in fanning interest in the problem with an idea that remains a viable possibility even to this day.

Paul Dirac was the Lucasian Professor of Mathematics at Cambridge for some of the time when Eddington was living at the Observatories. Stories of Dirac's simple and entirely logical approach to life and social interaction are legion and it is entirely in keeping with their peculiar tenor to find that his unexpected foray into the issue of Large Numbers was written whilst on his honeymoon, in February 1937.[7]

Unpersuaded by Eddington's numerological approach to the presence of 'large numbers' amongst the constants of Nature,[8] Dirac argued that very large dimensionless numbers taking values like 10^{40} and 10^{80} are most unlikely to be independent and unrelated accidents: there must exist some undiscovered mathematical formula linking the quantities involved. They must be consequences rather than coincidences. This is Dirac's Large Numbers Hypothesis (LNH):

> 'Any two of the very large dimensionless numbers occurring in Nature are connected by a simple mathematical relation, in which the coefficients are of the order of unity.'[9]

The large numbers that Dirac marshalled to motivate this daring new hypothesis drew on Eddington's work and were three in number:

$$N_1 = \text{(size of the observable universe)}/\text{(electron radius)}$$
$$= ct/(e^2/m_e c^2) \approx 10^{40}$$

$$N_2 = \text{electromagnetic-to-gravitational force ratio between}$$
$$\text{proton and electron}$$
$$= e^2/Gm_e m_{pr}) \approx 10^{40}$$

$$N = \text{number of protons in the observable universe}$$
$$= c^3 t/Gm_{pr} \approx 10^{80}$$

Here, t is the present age of the Universe, m_e is the mass of an electron, m_{pr} is the proton mass, G the constant of gravitation, c the speed of light, and e the electron charge.

According to Dirac's hypothesis, the numbers N_1, N_2 and $\sqrt{N}$ were actually *equal* up to small numerical factors of order unity. By this, he meant that there must be laws of Nature that require formulae like $N_1 = N_2$ or even $N_1 = 2N_2$. A number like 2, or 3, not terribly different from 1 is permitted because it is so much smaller than the large numbers involved in the formula; this is what he meant by the 'coefficients . . . of the order of unity.'

This hypothesis of equality between Large Numbers is not in itself original to Dirac. Eddington and others had written down such relations before, but Eddington had not distinguished between the number of particles in the entire universe – which might be infinite – and the number of particles in the observable Universe, which is defined to be a sphere about us with radius equal to the speed of light times the present age of the Universe. The radical change precipitated by Dirac's LNH is that it requires us to believe that *a collection of traditional constants of Nature, like N_2 must be changing as the universe ages in time, t*:

$$N_1 \approx N_2 \approx \sqrt{N} \propto t$$

Because Dirac had included two combinations which included the age of the Universe, *t*, in his catalogue of Large Numbers, the relation he proposes requires that a combination of three of the traditional constants of Nature is not constant at all but must increase steadily in value as the Universe ages, so

$$e^2/Gm_{pr} \propto t \qquad (*)$$

Dirac chose to accommodate this requirement by abandoning the constancy of Newton's gravitation constant, G. He suggested that it was decreasing in direct proportion to the age of the Universe over cosmic time scales, as

$$G \propto 1/t$$

Thus in the past G was bigger and in the future it will be smaller than it is measured to be today. One now sees that $N_1 \approx N_2 \approx \sqrt{N} \propto t$ and the huge magnitude of the three Large Numbers is a consequence of the great age of the universe:[10] they all get larger as time goes on.[11]

Dirac's proposal provoked a stir amongst a group of vociferous scientists who filled the letters pages of the journal *Nature* with arguments for and against.[12] Meanwhile, Dirac maintained his customary low profile, but wrote about his belief in the importance of large numbers for our understanding of the universe in words that might easily have been written by Eddington, so closely do they mirror the philosophy of his ill-fated 'Fundamental Theory':

> 'Might it not be that all present events correspond to prop-
> erties of this large number [10^{40}], and, more generally, that
> the whole history of the universe corresponds to proper-
> ties of the whole sequence of natural numbers . . . ? There
> is thus a possibility that the ancient dream of philosophers
> to connect all Nature with the properties of whole numbers
> will some day be realised.'[13]

Dirac's approach has two significant elements. First, he seeks to show that what might previously have been regarded as coincidences are consequences of a deeper set of relationships that have been missed. Second, he sacrifices the constancy of the oldest known constant of Nature. Unfortunately, Dirac's hypothesis did not survive for long. The proposed change in the value of G was just too dramatic. In the past gravity would have been much stronger; the energy output of the Sun would be changed and the Earth would have been far hotter in the past than usually assumed.[14] In fact, as the American physicist Edward Teller showed[15] in 1948, the oceans would have been boiling in the pre-Cambrian era, 200–300 million years ago, and life as we know it would not have survived, yet geological evidence available then showed that life had existed on Earth for at least 500 million years. Teller, an Hungarian émigré, was a high-profile physicist who played an important role in the development of the hydrogen bomb. He and Stan Ulam at Los Alamos were the two individuals who came up with the key idea (discovered independently by Andrei Sakharov in the Soviet Union) that showed how a nuclear bomb could be detonated. Later, Teller played a controversial role in the trial of Robert Oppenheimer and became an extreme hawk during the cold war period. He was the model for the character of Dr Strangelove so memorably portrayed by Peter Sellers in the film of that name. He is still an influential figure in weapons science and energy studies in the United States.

The exuberant George Gamow was a good friend of Teller's and responded to the boiling-ocean problem by suggesting that it could be ameliorated if it was assumed that Dirac's coincidences were created by a time variation in e, the electron charge, with e^2 increasing with time as the equation (*) on p. 101 requires.[16]

This suggestion didn't survive for long either. Unfortunately, Gamow's proposal for varying e had all sorts of unacceptable consequences for life on Earth. It was soon realised that Gamow's theory would have resulted in the Sun exhausting all its nuclear fuel long ago. The Sun would not be shining today if e^2 grows in proportion to the age of the universe. It would have been too small in the past to allow stars like the Sun to form.

Gamow had a number of discussions with Dirac about these variants to his hypothesis of varying G. Dirac made an interesting response to Gamow with regard to his idea that the charge on the electron, and hence the fine structure constant, might be varying. No doubt recalling Eddington's early belief that the fine structure constant was a rational number, he writes to Gamow in 1961 about the cosmological consequences of its variation as the logarithm of the age of the Universe, that

> 'It is difficult to make any firm theories about the early stages of the universe because we do not know whether hc/e^2 is a constant or varies proportional to $log(t)$. If hc/e^2 were an integer it would have to be a constant, but the experimenters now say it is not an integer, so it might very well be varying. If it does vary, the chemistry of the early stages would be quite different, and radio-activity would also be affected. When I started work on gravitation I hoped to find some connection between it and neutrinos, but this has failed.'[17]

Dirac was not readily going to subscribe to varying e as a solution of the Large Numbers conundrum. His most important scientific work had involved understanding the structure of atoms and the behaviour of the electron. All this was based upon the assumption, shared by everyone else, that e was a true constant, the same at all times and in all places in the Universe. Even Gamow soon gave up his theory about varying e and concluded that

> 'The value of e stands as the Rock of Gibraltar for the last 6×10^9 years!'[18]

Dirac's suggestion attracted wide interest from scientists of many unexpected quarters. Alan Turing, a pioneer in cryptography and the theory of computation, was fascinated by the idea of changing gravity and speculated about whether it would be possible to test the idea from fossil evidence, asking if

> 'a paleontologist could tell from the footprint of an extinct animal, whether its weight was what it was supposed to be.'[19]

The great biologist J.B.S. Haldane[20] became fascinated by the possible biological consequences of cosmological theories in which traditional 'constants' change with time or where gravitational processes unfolded with respect to a different cosmic clock than atomic processes. Such two-timing universes had been proposed by Milne and were the first suggestions that G might not be constant. Processes, like radioactive decay or molecular interaction rates, might be constant on one timescale but significantly variable with respect to the other. This gave rise to a scenario in which life-sustaining biochemistry only became possible after a particular cosmic epoch. Haldane suggests that

> 'There was, in fact, a moment when life of any sort first became possible and the higher forms of life may only have

become possible at a later date. Similarly a change in the properties of matter may account for some of the peculiarities of pre-Cambrian geology.'

This imaginative scenario is not dissimilar to that now known as 'punctuated equilibrium' in which evolution occurs in a staccato succession of accelerated bouts interspersed by long periods of slow change. However, Haldane provides an explanation for the changes.

What all these diverse responses to the ideas of Eddington and Dirac have in common is a growing appreciation that constants of Nature play a vital cosmological role: that there is a link between the structure of the Universe as a whole and the local conditions within it that are necessary for life to evolve and persist. If the traditional constants vary then astronomical theories have big consequences for biology, geology and life itself.

OF THINGS TO COME AT LARGE

'The baby figure of the giant mass
Of things to come at large.'

William Shakespeare, *Troilus and Cressida*

The short-term legacy of the early interest in large numbers involving the constants of Nature was a focus of interest upon the possibility that some traditional constants of Nature might be varying very slowly over the billions of years of cosmic history. New theories of gravity were developed, extending Einstein's general theory of relativity to include varying gravity. Instead of being treated as a constant, Newton's G was like temperature, able to vary in strength from place to place and with the passage of time. Fortunately, this is not as hopelessly unconstrained as it might at first sound. In order that the changes in G respect the laws of cause and effect, not propagate changes at speeds faster than light, and don't violate the conservation of energy, there is a single type of theory which fits the bill. Many scientists found parts of this theory but the simplest and most complete representation of it was written down by the American physicist Robert Dicke and his research student, Carl Brans, in 1961.

Dicke was a rare physicist. He was equally at home as a mathematician, an experimental physicist, a distiller of complicated astronomical data, or the designer of sophisticated measuring instruments. He had the widest possible scientific interests. He realised that the idea of a varying 'constant' of gravity could be subjected to a plethora of observational tests, using the data of geology, palaeontology, astronomy and laboratory physics. Nor was he motivated simply by a desire to explain the Large Numbers. During the mid 1960s there was a further motivation for developing an extension of Einstein's theory of gravity that included varying G. For a while it appeared that Einstein's predictions about the wobble in the orbit of the planet Mercury did not agree with observations when the slightly non-spherical shape of the Sun was taken into account.

Dicke showed that if a variation in G with time was allowed then its rate of change could be chosen to have a value that would agree with the observations of Mercury's orbit. Sadly, years later this was all found to be something of a wild goose chase. The disagreement with Einstein's theory was being created by inaccuracies in our attempts to measure the diameter of the Sun which made it seem that the Sun was a different shape than it really was. The Sun's size is not so easy to measure at the levels of accuracy required because of the turbulent activity on the solar surface. When this problem was resolved in 1977 the need for a varying G to reconcile the observations with theory disappeared.[21]

In 1957, whilst beginning to develop theories with varying G, Dicke prepared a major review about the geophysical, palaeontological and astronomical evidences for possible variations of the traditional physical constants. He made the interesting remark that the issue of explaining the 'large numbers' of Eddington and Dirac must have some biological aspect:[22]

> 'The problem of the large size of these numbers now has a ready explanation . . . there is a single large dimensionless number which is statistical in origin. This is the number of particles in the Universe. The age of the Universe "now" is not random but is conditioned by biological factors . . . [because changes in the values of Large Numbers] would preclude the existence of man to consider the problem.'

Four years later, he elaborated this important insight in more detail, with special reference to Dirac's Large Number coincidences, in a short letter published in the journal *Nature*. Dicke argued that biochemical life-forms like ourselves owe their chemical basis to such elements as carbon, nitrogen, oxygen and phosphorus which are synthesised after billions of years of main-sequence stellar evolution. (The argument applies with equal force to any life-form based upon any atomic elements heavier than helium.) When stars die these 'heavy' biological elements are dispersed throughout space by supernovae from whence they are incorporated into grains, planetesimals, planets, self-replicating 'smart' molecules like DNA and, ultimately, into ourselves. Observers cannot arise until roughly the hydrogen-burning lifetime of a main-sequence star has elapsed and it is difficult for them to survive after the stars have burnt out. This time scale is controlled by fundamental constants of Nature to be

$$t(star) \approx (Gm_{pr}^2/hc)^{-1} \, h/m_{pr}c^2 \approx 10^{40} \times 10^{-23} \text{ seconds} \approx 10 \text{ billion years}$$

We would not expect to be observing the Universe at times significantly in excess of *t(star)*, since all stable stars would have expanded, cooled and died. Nor would we be able to see the Universe at times much less than $t(star)$ because we could not exist! There would be neither stars nor heavy elements like carbon. We seem strait-jacketed by the facts of biological life to gaze at the Universe and develop cosmological theories after a time $t(star)$ has elapsed since the Big Bang. Thus the value of Dirac's Large Number, $N(t)$ is by no means random. It must have a value close to the value taken by $N(t)$ when t is close in value to $t(star)$.

If we look at the value of N at the time $t(star)$ we find that it is precisely Dirac's Large Number coincidence. All Dirac's coincidence is saying is that we live at a time in cosmic history after the stars have formed and before they die. This is not surprising. Dicke is telling us that we could not fail to observe Dirac's coincidence: it is a prerequisite for life of our sort to exist. There is no need to give up Einstein's theory of gravitation by requiring G to vary, as Dirac implicitly required, nor do we need to deduce some numerological connection between the strength of gravity and the number of particles in the universe as Eddington had thought. The Large Number coincidence is no more surprising than the existence of life itself.

Dirac's response, his first written remarks about cosmology for more than twenty years, to this unusual perspective upon cosmological observations was rather bland:

> 'On Dicke's assumption habitable planets could exist only for a limited period of time. With my assumption they could exist indefinitely in the future and life need never end. There is no decisive argument for deciding between these assumptions. I prefer the one that allows the possibility of endless life.'

Although he was willing to admit that life would be unlikely to exist before the stars had formed he was unwilling to concede that it could not continue long after they had burnt out. With Dirac's idea of varying G the coincidences would continue to be seen at all times but on

Dicke's hypothesis they would only be seen near the present epoch. Dirac didn't think there was any problem with having habitable planets in the far future on his theory. However, if gravity is getting weaker it is not clear that stars and planets would be able to exist in the far future. At the very least, other constants would need to vary to maintain the balance between gravity and the other forces of Nature that make their existence possible.

It is very striking that other notable cosmologists like Milne had previously argued in the opposite way to Dicke. Milne regarded the appearance of Large Number coincidences in Eddington's theories as suspicious. He didn't believe that any 'Fundamental Theory' of Nature could possibly hope to explain coincidences between large numbers precisely because the large numbers involved the present age of the Universe. Since there was nothing special about the present time we were living at, no fundamental theory of physics could predict it or pick it out and so it could not explain the coincidences:

> 'There is necessarily an empirically defined quantity, t [the present age of the universe] occurring in these expressions, for this simply measures the position of the instant at which we happen to be viewing the universe. This, of course, is incapable of prediction . . . The circumstance that Eddington's theory of the constants of nature appears to predict this . . . on *a priori* grounds seems to me an argument against Eddington's theory . . . for it appears to be equivalent to the feat of predicting the age of the universe at the moment we happen to be viewing it; which would be absurd.'[23]

Dicke showed that, on the contrary, you certainly could predict something very particular about the age of the Universe if carbon-based beings are doing the predicting.

Dicke's point can be restated in an even more striking fashion. In order for a Big Bang universe to contain the basic building blocks[24] necessary for the subsequent evolution of biochemical complexity it must have an age at least as long as the time it takes for the nuclear reactions in stars to produce these elements. This means that the observable Universe must be at least ten billion years old and so, since it is expanding, it must be at least ten billion light years in size. We could not exist in a universe that was significantly smaller.

Despite Dirac's dislike of Dicke's approach, we can find an unusual application of a similar idea being introduced by him to save his theory that G falls as the universe ages. After Edward Teller and others had discovered the problems that such a radical variation of gravity would create for the history of stars and life on Earth, there were attempts to keep the varying-G theory alive by assuming that stars like the Sun periodically passed through dense clouds from which they accreted material fast enough to offset the effects of decreasing G on the Sun's gravitational pull. Gamow thought that such an assumption

> 'would be extremely *un*elegant, so that the total amount of elegance in the entire theory would have decreased quite considerably even though the elegant assumption [$G \propto t^{-1}$] would be saved. So, we are thrown back to the hypothesis that 10^{40} is simply the largest number the almighty God could write during the first day of creation.'

It is interesting to note Gamow's stress on the 'inelegance' of fudging the theory in this way, since Dirac always urged others to look for 'beauty' (which is not necessarily the same thing as simplicity, he liked to point out) in the equations describing a physical theory. Indeed, he once wrote to Heisenberg about one of his proposed theories that

> 'My main objection to your work is that I do not think your basic . . . equation has sufficient mathematical beauty to be a fundamental equation of physics. The correct

equation, when it is discovered, will probably involve some new kind of mathematics and will excite great interest among pure mathematicians.'[25]

Yet Dirac was happy to defend the accretion idea, no matter how improbable it might appear, on the ground that it might be necessary for life to exist:

> 'I do not see your objection to the accretion hypothesis. We may assume that the sun has passed through some dense clouds, sufficiently dense for it to pick up enough matter to keep the earth at a habitable temperature for 10^9 years. You may say that it is improbable that the density should be just right for this purpose. I agree. *It is improbable*. But this kind of improbability does not matter. If we consider all the stars that have planets, only a very small fraction of them will have passed through clouds of the right density to maintain their planets at an equable temperature long enough for advanced life to develop. There will not be so many planets with men on them as we previously thought. However, provided there is one, it is sufficient to fit the facts. So there is no objection to assuming our sun has had a very unusual and improbable history.'[26]

This is a remarkable about-face, six years after his initial opposition to Dicke's inclusion of human life as a factor in assessing the likelihood of unusual situations arising.

A beautifully simple argument regarding the inevitability of the large size of the Universe *for us* appears first in the text of the Bampton Lectures given by the Oxford theologian Eric Mascall. They were published in 1956 under the title *Christian Theology and Natural Science* and he attributes the basic idea to Gerald Whitrow. Stimulated by Whitrow's suggestions, he writes:

'if we are inclined to be intimidated by the mere size of the Universe, it is well to remember that on certain cosmological theories there is a direct connection between the quantity of matter in the Universe and the conditions in any limited portion of it, so that in fact it may be necessary for the Universe to have the enormous size and complexity which modern astronomy has revealed, in order for the earth to be a possible habitation for living beings.'[27]

This simple observation can be extended to provide us with a profound understanding of the subtle links that exist between superficially different aspects of the Universe we see around us and the properties that are needed if a Universe is to contain living beings of any sort.

BIG AND OLD, DARK AND COLD

'It's a funny old world – a man's lucky if he gets out of it alive.'

W. C. Fields[28]

We have seen that the process of stellar alchemy takes time – billions of years of it. And because our Universe is expanding it needs to be billions of light years in size if it is to have enough time to produce the building blocks for living complexity. A universe that was only as big as our Milky Way galaxy, with its 100 billion stars, would be little more than a month old. Another consequence of an old expanding universe, besides its large size, is that it is cold, dark and lonely. When any ball of gas or radiation is expanded in volume, the temperature of its constituents falls off in proportion to the increase in its size. A universe that is big and old enough to contain the building blocks of complexity will be very cold and the levels of average radiant energy so low that space will everywhere appear dark.

It is sobering to reflect upon all the metaphysical and religious responses there have been, over the centuries, to the darkness of the night sky and the patterns of stars embroidered upon it; to the vastness of space and our incidental place within it, a mere dot in the grand scheme of things. Modern cosmology shows that these features are not random accidents. They are part and parcel of the whole interconnectedness of the universe. They are, in fact, necessary features of any universe that contains living observers. Remarkably, the metaphysical effect of this type of universe upon its inhabitants may well be another inescapable by-product for any sentient beings elsewhere as well. The Universe has the curious property of making living beings think that its unusual properties are unsympathetic to the existence of life when in fact they are essential for it.

If we were to smooth out all the material in the Universe into a uniform sea of atoms we would see just how little of anything there is. There would be little more than about 1 atom in every cubic metre of space. No laboratory on Earth could produce an artificial vacuum that was anywhere near as empty as that. The best vacuum achievable today contains approximately 1000 billion atoms in a cubic metre.

This way of looking at the Universe provides some important new insights into the properties it displays to us. Many of its most striking features — its vast size and huge age, the loneliness and darkness of space — are all necessary conditions for there to be intelligent observers like ourselves. We should not be surprised that extraterrestrial life, if it exists, is so rare and so far away. The low average density of matter in the Universe means that if we were to aggregate material into stars or galaxies, then we should expect huge distances to lie between these objects on average. In Figure 6.1, the density of material in the Universe is expressed in a variety of different ways which shows how far apart we should expect planets, stars and galaxies to be.

In Figure 6.2 we show the expanding trajectory of our Universe as time passes. Gradually, the environment within in the Universe cools

---

The Visible Universe contains
only

- 1 Atom per cubic metre
- 1 Earth per $(10 \text{ light-yr})^3$
- 1 Star per $(10^3 \text{ light-yr})^3$
- 1 Galaxy per $(10^7 \text{ light-yr})^3$
- 1 'Universe' per $(10^{10} \text{ light-yr})^3$

Figure 6.1 *The density of matter in our Universe expressed in a number of different units of volume that show how rare galaxies, stars, planets, and atoms actually are on the average. We should not be surprised to find that extraterrestrial life is very rare.*

off and allows atoms, molecules, galaxies, stars and planets to form. We are located in a particular niche of cosmic history between the birth and death of the stars.

It appears that the existentialist philosopher Karl Jaspers was also provoked by Eddington's writings to consider the significance of our existence in a particular locale at a particular epoch of cosmic history. In his influential book,[29] written in 1949, soon after Eddington's death, he asks

'Why do we live and accomplish our history in infinite

Figure 6.2 *The changing environment in an expanding universe like our own. As the universe cools and ages it is possible for atoms, molecules, galaxies, stars, planets, and living organisms to form. In the future the stars will extinguish their nuclear fuel and die. There is a niche of cosmic history in which our sort of biological evolution must first occur if it is ever to occur.*

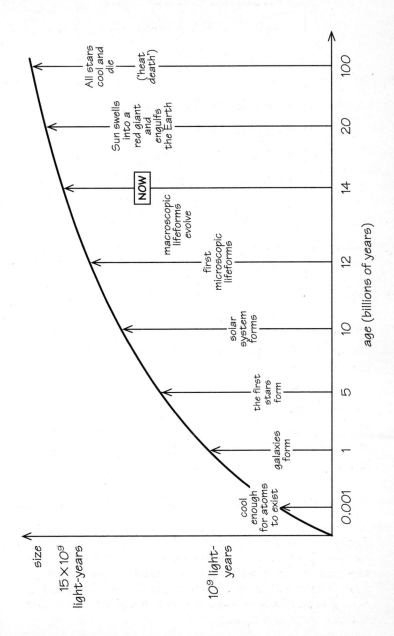

space at precisely this point, on a minute grain of dust in the universe, as though in an out-of-the-way corner? Why just now in infinite time? These are questions whose unanswerability makes us conscious of an enigma.

The fundamental fact of our existence is that we appear to be isolated in the cosmos. We are the only articulate rational beings in the silence of the universe. In the history of the solar system there has arisen on the earth, for a so far infinitesimally short period, a condition in which humans evolve and realise knowledge of themselves and of being . . . Within the boundless cosmos, on a tiny planet, for a tiny period of a few millennia, something has taken place as though this planet were the all-embracing, the authentic. This is the place, a mote in the immensity of the cosmos, at which being has awakened with man.'

There are some big assumptions here about the uniqueness of human life in the Universe. Yet, the question is raised, although not answered, as to why we are here at the time and place that we are. We have seen that modern cosmology can provide an illuminating response to this question.

THE BIGGEST NUMBER OF ALL

'Al-Gore-rithm, *n*. a mathematical operation which is repeated many times until it converges to the desired result, especially in Florida.'

The Grapevine

Astronomers are used to huge numbers. They are challenged to explain to outsiders just what billions and billions of stars really means with some homespun analogy. It was only when the American national debt grew to astronomical levels that there were suddenly numbers in the financial pages of newspapers that were larger than the number of stars in the Milky Way or galaxies in the Universe.[30] Yet, curiously, if you want really big numbers, numbers that dwarf even the 10^{80}s of Eddington and Dirac, astronomy is not the place to look. The big numbers of astronomy are additive. They arise because we are counting stars, planets, atoms and photons in a huge volume. If you want really huge numbers you need to find a place where the possibilities multiply rather than add. For this you need complexity. And for complexity you need biology.

In the seventeenth century the English physicist Robert Hooke made a calculation 'of the number of separate ideas the mind is capable of entertaining'.[31] The answer he got was 3,155,760,000. Large as this number might appear to be (you would not live long enough to count up to it!) it would now be seen as a staggering underestimate. Our brains contain about ten billion neurons, each of which sends out feelers, or axons, to link it to about one thousand others. These connections play some role in creating our thoughts and memories. How this is done is still one of Nature's closely guarded secrets. Mike Holderness suggests that one way of estimating[32] the number of possible thoughts that a brain could conceive is to count all those connections. The brain can do many things at once so we could view it as some number, say a thousand, little groups of neurons. If each neuron makes a thousand different links to the ten million others in the same group then the number of different ways in which it could make connections in the same neuron group is $10^7 \times 10^7 \times 10^7 \times \ldots$ one thousand times. This gives 10^{7000} possible patterns of connections. But this is just the number for one neuron group. The total number for 10^7 neurons is 10^{7000} multiplied together 10^7 times. This is $10^{70,000,000,000}$. If the 1000 or so groups of neurons can operate independently of each other then each of them contributes $10^{70,000,000,000}$ possible wirings, increasing the total to the Holderness number, $10^{70,000,000,000,000}$.

This is the modern estimate of the number of different electrical patterns that the brain could hold. In some sense it is the number of different possible thoughts or ideas that a human brain could have. We stress the '*could*'. This number is so vast that it dwarfs the number of atoms in the observable Universe – a mere 10^{80}. But unlike the number of atoms in the Universe it does not gain its vastness from filling up a huge volume with little things. The brain is rather small. It only contains about 10^{27} atoms. The huge number comes from the potential complexity of the number of connections between components. This is what we mean by complexity. It arises from the number of different ways in which components can be connected together, rather than out of the identity of those components. And, because these seriously large numbers arise out of the number of permutations available to a complex network of switches they will not be explainable in terms of the constants of Nature in the way that the astronomical Large Numbers are. They are not only bigger; they're also different.

chapter seven

Biology and the Stars

'Things are more like they are now than they ever were before.'

Dwight D. Eisenhower

IS THE UNIVERSE OLD?

'The four ages of Man: Lager, Aga, Saga, and Gaga.'

Anonymous

When we think about the age and the size of the Universe we generally do so using measures of time and space like years, kilometres or light years. As we have already seen, these measures are extremely anthropomorphic. Why measure the age of the Universe using a 'clock' that 'ticks' once every time our planet completes an orbit around its parent star? Why measure its density in terms of atoms in a cubic metre? The answers to these questions are of course the same: because it's convenient and we've always done it like that. But here is a situation where it is especially appropriate to use the 'natural' units of mass, length and time that Stoney and Planck introduced to help us escape from the strait-jacket of a human-centred perspective.

If we adopt Planck's units then we see that the

$$present\ age\ of\ visible\ universe \approx 10^{60}\ Planck\text{-}times$$

The size of the visible Universe is similarly huge:

$$present\ size\ of\ visible\ universe \approx 10^{60}\ Planck\text{-}lengths$$

and so is its mass:

$$present\ mass\ of\ visible\ universe \approx 10^{60}\ Planck\text{-}masses$$

Thus we see that the very low density of matter in the Universe is a reflection of the fact that

$$present\ density\ of\ visible\ universe \approx 10^{-120}\ of\ the\ Planck\ density$$

and the temperature of space, at three degree above absolute zero, is

$$present\ temperature\ of\ visible\ universe \approx 10^{-30}\ of\ the\ Planck\ temperature$$

These extremely large numbers and tiny fractions show us immediately that the Universe is structured on a superhuman scale of staggering proportions when weighed in the balances of its own construction. By its own standards the Universe is old. The natural lifetime of a world governed by gravity, relativity and quantum mechanics is the fleetingly brief Planck time. Somehow our Universe has managed to keep expanding for a vast number of Planck times. It seems to be much older than it should be. Later we shall see that cosmologists think they know how this came about. Yet, despite the huge age of the Universe in 'ticks' of Planck time, we have learnt that almost all this time is needed to produce stars and the life-supporting chemical elements.

THE CHANCE OF A LIFETIME

'At the End of the Universe you have to use the past tense a lot . . . everything's been done you know.'

Douglas Adams[1]

Why isn't our Universe much older than it appears to be? It is easy to understand why the Universe isn't much younger. Stars take a long time to form and to produce the heavier elements that biological complexity requires. But old universes have their problems as well. As time passes in the Universe the process of star formation will slow down. All the gas and dust that forms the raw material for stars will have been processed by stars and ejected into intergalactic space where it is unable to cool down and coalesce into new stars. Few stars means few solar systems and planets. Any planets that do form are less active than those formed earlier. The production of radioactive elements in the stars will diminish and those that are formed will have longer half-lives. New planets will be less geologically active and will lack many of the subterranean movements that power vulcanism, continental drift and mountain uplifting on the Earth. If this also makes the presence of a magnetic field less likely on a planet then life will be very unlikely to evolve into complex forms. Typical stars, like our Sun, emit a wind of electrically-charged particles from their surface which will strip off the atmospheres of orbiting planets unless the wind can be deflected by a planetary magnetic field. In our solar system the Earth's magnetic field has protected its atmosphere from the solar wind but Mars, unprotected by any magnetic field, lost its atmosphere long ago.

Long life on a solar system planet is probably not easy to sustain. We have gradually come to appreciate how precarious it is. Putting to one side the attempts that living beings persist in making to extinguish themselves, exhaust natural resources, spread lethal infections and deadly poisons, there are serious outside threats as well. The movements of comets and asteroids are a serious hazard to the development and persistence of intelligent life in its early stages. Impacts are not uncommon and have had catastrophic effects on Earth in the distant past. We are fortunate to be doubly shielded from these impacts – by our small near neighbour, the Moon, and by our giant distant neighbour, Jupiter. Jupiter is a thousand times more massive than the Earth and sits on the outskirts of the solar system where its powerful gravitational pull can capture incoming objects heading for the inner solar system. In July 1994 we were able to witness the fragmentation and capture of comet Schumacher-Levy 9 by Jupiter.[2] In the twentieth century we had two significant impacts on Earth, one in South America and the other in Tunguska in northern Russia. We have been cheating the law of averages and one day our luck will change. Some governments are already investing effort in monitoring asteroids and planning countermeasures against Earthbound objects. Clearly, the longer a planet is around the greater its chances of being hit (see Figure 7.1).

These outside interventions upon the evolution of the Earth have a curious flip-side. True, they can produce global extinctions and set back the evolution of complexity by millions of years. But, in moderation, they may have a positive, accelerating effect upon the evolution of intelligent forms of life. When the dinosaurs were extinguished by the impact of a large meteor or comet striking the Earth on the Yucatan peninsula 65 million years ago at the end of the Mesozoic Era, the Earth was rescued from an evolutionary dead end. The dinosaurs seem to have evolved along a track that developed physical size rather than brain size. The disappearance of the dinosaurs, along with most other life-forms on Earth at that time, opened up space for the emergence of mammals. It also cleared niches of competitors for natural resources.

Figure 7.1 *The average frequency of meteor impacts of different sizes on the Earth's atmosphere. Also shown is the diameter of the meteor and the diameter of the crater left on the Earth's surface together with the likely effects.*[3]

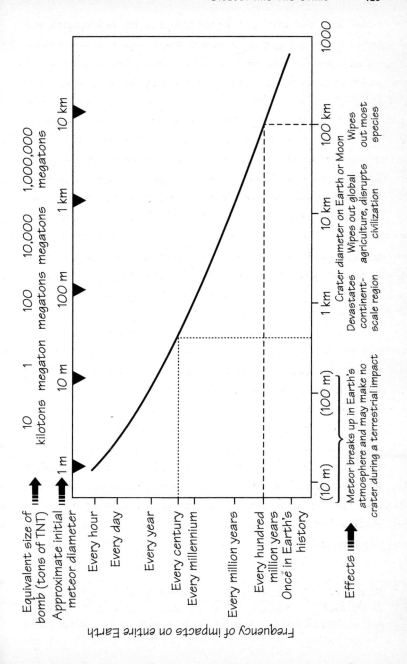

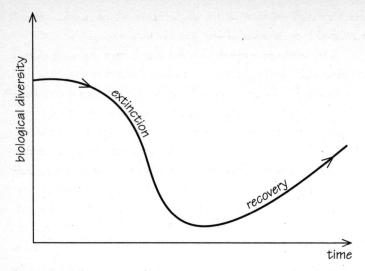

Figure 7.2 *The pattern of response to an environmental crisis that produces a mass extinction event on Earth.*[4]

This spurred a rapid acceleration in the development of diversity. Perhaps impacts play a vital role in kick-starting evolution when it gets stuck along unpromising paths. Without impacts the development process may settle into a stable, but unexciting byway with steady extinctions reducing species diversity constantly (see Figure 7.2). Harsh, fast-changing conditions stimulate adaptation and accelerate the evolutionary process. They also increase diversity and the best life insurance a planet can take out against total extinction of its biology by a future impact is to create diversity. You won't see it that way if you're a dinosaur though.

In our solar system life first evolved surprisingly soon after the formation of a hospitable terrestrial environment. There is something unusual about this. Suppose the typical time that it takes for life to evolve is called $t(bio)$, then from the evidence of our solar system, which is about 4.6 billion years old, it seems that the time it takes for stars to settle down and create a stable source of heat and light, $t(star)$, is not very different to $t(bio)$ because we have found simple forms of terrestrial bacterial life that are several billion years old.

This similarity between $t(bio)$ and $t(star)$ seems like a coincidence. At first sight we might assume that the microscopic biochemical processes and local environmental conditions that combine to determine the magnitude of $t(bio)$ are *independent* of the nuclear astrophysical and gravitational processes that determine the typical stellar lifetime of a star. However, this assumption leads to the striking conclusion that we should expect extraterrestrial forms of life to be exceptionally rare. The argument, in its simplest form, introduced by Brandon Carter[5] and developed further by myself and Frank Tipler,[6] and still investigated minutely today[7] goes like this. If $t(bio)$ and $t(star)$ are unconnected with one another then the time that life takes to arise will be random with respect to the stellar time scale $t(star)$. So it is most likely[8] that we should either find that $t(bio)$ is much bigger than $t(star)$ or find that $t(bio)$ is much less than $t(star)$.

Now let's take stock. On the one hand, if $t(bio)$ is typically much less than $t(star)$ we need to ask why it is that the first observed inhabited solar system (ours!) has $t(bio)$ approximately equal to t(star). By our logic, this would be extraordinarily unlikely. On the other hand, if $t(bio)$ is typically much greater than $t(star)$ then the first observed inhabited solar system (ours) is a statistical fluke and most likely to have $t(bio)$ approximately equal to $t(star)$, since systems with $t(bio)$ much greater than $t(star)$ have yet to evolve. Thus we are led to conclude that we are a rarity, one of the first living systems to arrive on the scene.

In order to escape from this conclusion we have to undermine one of the assumptions underlying the argument that leads to it. For example, if we suppose that $t(bio)$ is *not* independent of $t(star)$, then things look different. If the ratio $t(bio)/t(star)$ increases with $t(star)$ then it may become likely that we will find $t(bio)$ approximately equal to $t(star)$. Mario Livio[9] has suggested how $t(bio)$ and $t(star)$ could be related by a relation of this general form if the evolution of a life-supporting planetary atmosphere requires an initial phase during which oxygen is released by the photodissociation of water vapour. On Earth this took 2.4 billion years and built up the atmospheric oxygen to about one thousandth of its present value. The length of this phase might be expected to be inversely proportional to the intensity of radiation in the wavelength interval 1000–2000 Ångstroms, where the key molecular levels for water absorption lie. Further studies of stellar evolution may allow us to determine the length of this phase and so establish a link between the biological evolution time and the main-sequence stellar lifetime.

This model indicates a possible route to establishing a link between the biochemical time scales for the evolution of life and the astrophysical time scales that determine the time required to create an environment supported by a stable hydrogen-burning star. There are obvious weak links in the argument, though. It provides only a necessary condition for life to evolve, not a sufficient one. We could imagine an expression for the probability of planet formation around a star. It would involve many other factors which would determine the amount of material available for the formation of solid planets with atmospheres at distances which permit the presence of liquid water and stable surface conditions. Furthermore, we know that there were many 'accidents' of the planetary formation process in the solar system which have subsequently played a major role in the existence of long-lived stable conditions on Earth. For example, as Jacques Laskar and his co-workers[10] have shown, the presence of resonances between the precession rates of rotating planets and the gravitational perturbations they feel from all other bodies in their solar system can easily produce chaotic evolution of the tilt of a planet's rotation axis with respect to the orbital plane of the planets over times much shorter than the age of the solar system. The planet's surface temperature variations and sea levels are sensitive to this angle of tilt. It determines the climatic differences between what we call 'the seasons'. In the case of the Earth, the modest angle of tilt (approximately 23 degrees) would have experienced this erratic evolution had it not been for the presence of the

Moon. The Moon is so large that its gravitational effects dominate the resonances which exist between the Earth's precessional rotation and the frequency of external gravitational perturbations from the other planets. As a result the Earth's tilt wobbles only by half a degree around 23 degrees over hundreds of thousands of years.

This shows how the causal link between stellar lifetimes and biological evolution times may be rather a minor factor in the chain of fortuitous circumstances that must occur if habitable planets are to form and sustain viable conditions for the evolution of life over long periods of time.

OTHER TYPES OF LIFE

'Life is not for everyone.'

Michael O'Donoghue[11]

One of the assumptions that arguments for the inevitability of a large and cool universe are implicitly making is that all life is very much like us. Biologists seem happy to admit the possibility of other forms of life but are not so sure that it is likely to evolve spontaneously, without a helping hand from carbon-based life-forms. Most estimates of the likelihood of extraterrestrial intelligences existing in the Universe focus upon life-forms similar to ourselves who live on planets and need water, gaseous atmospheres and so forth. It is worth stretching our imaginations a little to think about what life might be like if it was space-based rather than planet-based. The astronomer Fred Hoyle created an interesting example which he hoped would evade the usual unfavourable conclusions about likelihood that had been made for planet-based ETIs. Not content with successful careers as astrophysicist and populariser of science, Hoyle branched out into science fiction, with notable successes. His most famous story, *The Black Cloud*,[12] was a big publishing success that created a plausible contemporary thriller involving scientists not dissimilar to Hoyle himself. Indeed, despite his assurances that the characters are entirely fictitious, it is hard not to identify the hero with Hoyle himself. *The Black Cloud* was written in 1957, just a few years after the discovery of coincidences concerning the values of the constants of Nature that have important implications for the possible existence of carbon and oxygen, and hence for life in the Universe. There was much discussion about the likelihood of life elsewhere and the first two Soviet *Sputnik* space probes were launched in 1957. The scenario is set for Earth to face the approach from space of a cloud of gas, of which there are many in interstellar space, which is on course to pass between the Earth and the Sun. If it does then the heat and light from the Sun will be cut off for a period, after being amplified for a while by reflection from the cloud, with potentially calamitous consequences for Earth. Events take an unexpected turn. The cloud turns out to be intelligent, an amorphous life-form existing as a huge system of complex molecular correlations moving through space. After much intrigue and excitement the Earth survives its brief encounter with the passing cloud but not before it has established a dialogue with it and learned to decode the signals it uses to speak to us. Yet the most important message that Hoyle was trying to get across in this story was the possible error of assuming that life lives on solid planets. Perhaps the chemical complexity needed to qualify as 'life' could exist in huge molecular clouds, stabilised by the binding force of gravity. Even carbon might not be needed in these nebulous cradles of life. Thirty years later Hoyle was to return to this theme in his scientific work and science fiction, imagining that self-reproducing molecules could have evolved within cometary interiors and then spread around the galaxies by the motion of the comets.

Other science fiction writers had explored the possibilities of alternatives to carbon chemistry. Silicon was known to form chain molecules a little like carbon does, but unfortunately they tend to be like quartz and sand, rigid and uninteresting as a building block for biology.

Ironically, the computer revolution has since shown that it is silicon *physics* rather than silicon chemistry which holds the most promise as another basis for life. But such artificial forms of life and intelligence are not spontaneously evolved. They have required the help of carbon-based organisms to bring together the highly organised, and hence extremely improbable, configurations needed for their persistence and development. These more abstract alternatives to life in flesh and blood form are now rather familiar to us and science fiction writers have to be considerably more subtle than merely to imagine aliens with odd chemistries and new bodily forms. But back in 1957 Hoyle's idea was a novel one. It played an important role in widening the spectrum of possibilities for life beyond what most astronomers had in mind. The probability of life was not to be determined solely from the statistics of habitable planets with temperate atmospheres and surface water in orbit around friendly stars.

PREPARE TO MEET THY DOOM

'If you're killed, you've lost a very important part of your life.'

Brooke Shields[13]

There is one further curiosity about the coincidence that exists between the biological evolution time and astronomy. Since it is unsurprising that the ages of typical stars are similar to the present age of the Universe there is also an apparent coincidence between the age of the Universe and the time it has taken to evolve life-forms like ourselves. If we look back at how long our intelligent ancestors (*Homo sapiens*) have been on the scene we find that it is only about 200,000 years, which is much less than the age of the Universe, 13 billion years. Our human history has lasted for less than two hundred thousandths of the history of the Universe. But if our descendants could go on indefinitely into the future the situation for them would become very different. Suppose they were still thinking about these questions when the Universe was one thousand billion years old. Then they would calculate that their intelligent ancestors had been around for 1000 billion minus 13 billion plus 200,000 years. The answer 987.2 billion years is very similar to 1000 billion years. Our descendants would not think that the history of their civilisation lasted for just a tiny fraction of the history of the universe. Brandon Carter and Richard Gott have argued that this appears to make us rather special compared with observers in the far distant future. If you believe that our location in cosmic history should not be special in this way then you are led to a dramatic conclusion. In order to make sure that we and our descendants in the near future do not have a special view of cosmic history, thinking our own history is vastly less than the total history of the Universe, *we need to have no far future descendants*. If life on Earth disappeared in a few thousand years then all our descendants would observe roughly the same number for the fraction of cosmic history that has seen human civil-isation exist. Gott estimated that by this argument we should be 95 per cent confident that life on Earth will end between 5000 and 7.8 million years in the future.

There is no reason to confine this argument to such cataclysmic events as the extinction of human life. It is based upon the simple statistical fact that if you observe something at a random time there is a 95 per cent chance that you will be observing it during the middle 95 per cent of the period when it can be observed.[14] To show the versa-tility of this simple piece of statistics, Gott was asked to prepare a series of predictions for the January 1st, 2000 issue of the *Wall Street Journal*. The ones chosen are shown in Figure 7.3.

It is easy to work out these sorts of statistics for the precarious things of your choice. If the present time is to be random with respect to the total time over which something is observable then with 95 per cent confidence its future is expected to lie within a time interval bigger than 1/39th and 39 times its past age. If we only want 50 per cent confidence then its future will extend between one third and three times its past age.

Phenomenon and its starting date	will probably survive	
	more than	but less than
Stonehenge (2000 BCE)	102.5 years	156,000 years
Pantheon (126 CE)	48 years	73,086 years
Humans (*Homo Sapiens*) (200,000 years old)	5,100 years	7.8 million years
Great Wall (of China) (210 BCE)	56 years	86,150 years
Internet (1969)	9 months	1,209 years
Microsoft (1975)	7 months	975 years
General Motors (1908)	2.3 years	3,588 years
Christianity (c. 33 CE)	50 years	76,713 years
United States (1776)	5.7 years	8,736 years
New York Stock Exchange (1792)	5.2 years	8,112 years
Manhattan (purchased in 1626)	9.5 years	14,586 years
Wall Street Journal (1889)	2.8 years	4,329 years
New York Times (1851)	3.8 years	5,811 years
Oxford University (1249)	19 years	29,289 years

Figure 7.3 *With 95 per cent confidence these are the shortest and longest times that we expect the following structures and organisations to have lasted for or to last for in the future according to Richard Gott's predictions[15] on New Year's Day 2000.*

FROM COINCIDENCE TO CONSEQUENCE

'Moriarty: "All that I have to say has already crossed your mind."
Holmes: "Then possibly my answer has crossed yours."'

A. Conan Doyle[16]

Dicke's response to the problem of the Large Numbers had many important consequences. He showed that the approaches of Eddington and Dirac had been extreme and unwarranted. They had tried to explain the Large Number coincidences by making major changes to our theories of physics. Eddington wanted to create an ambitious new fundamental 'theory of everything' from which he imagined would flow new equations linking the constants of Nature in unsuspected ways, showing the Large Number coincidences to be consequences of a deep-laid scheme of Nature. Likewise, Dirac responded by giving up the constancy of one of the traditional constants of Nature, G, so as to allow the coincidences between different large numbers to be consequences of an as yet unknown theory of gravity and atomic phenomena. Dicke, by contrast, took a less iconoclastic approach. He recognised that not all moments of time are equal: we should only expect to be looking at the Universe when it is old enough for living beings to exist within it. As a result there is an irremovable bias besetting our astronomical observations that we do well to be aware of. This bias ensures that Dirac's coincidence between different Large Numbers will be observed by beings like ourselves. Dicke's lesson for scientists is a powerful and simple one and if you don't take it on board then, like Dirac and Eddington, you may be doomed to embark upon an unwarranted wild-goose chase, giving up well-established theories for speculative new possibilities. Critics who have not understood Dicke's contribution sometimes object it is 'not a scientific theory' because it makes no predictions and so 'cannot be tested'.

This is a serious misunderstanding. The recognition of observer bias is not somehow a rival scientific theory that needs to be tested. It is a principle of scientific methodology of which we remain unaware or wilfully ignore at our peril. It is just a sophisticated version of a principle that experimental scientists are very familiar with – experimental bias.

When you carry out an experiment or seek to draw conclusions from observational data the most important insight that the experimenter requires is the awareness of what biases beset the experiment. Such biases will make it easier to gather certain sorts of evidence than others and produce a misleading result. An interesting case that came to light in the newspapers was in regard to the controversial issue of mathematical achievement levels in tests by school children in different countries. For many years it had been claimed that the average achievement by pupils in some South-East Asian countries was significantly higher than in the United Kingdom. Then it came to light that the weakest pupils in that country were removed from the total who were evaluated at an earlier stage in the educational process. Clearly, the effect of their removal is to skew the average attainments to be higher than they would otherwise be. Another recent example that caught my eye was an American survey to discover if people who attended church also tended to have better health. A serious bias beset the final results because people who were extremely sick would be unable to attend church.

What these examples show is that scientists of all sorts must strive to be aware of any bias that might skew their data to produce a conclusion that is not present in the underlying evidence. Dicke noticed something similar in the astronomers' view of the Universe. Ignore the lesson of observer selection and wrong conclusions will be drawn.

The challenge of the Large Numbers played an important role in the development of our efforts to understand the structure of the

Universe and the range of possibilities available for the constants of Nature that supply the skeleton on which the outcomes of Nature's laws are fleshed out. It encouraged a serious questioning of the constancy of the traditional constants of Nature, especially Newton's 'constant' G, and led to the formulation of new theories of gravity which enlarged Einstein's theory to include this possibility. This also precipitated a broad change of outlook. Suddenly, subjects like biology and geology which had traditionally had very little to do with astronomy and cosmology, were seen to be of cosmic importance. A broadened perspective to cosmological thinking appeared. Some cosmological theories might be tested by geophysical or palaeontological evidence or lead to histories in which the evolution of life by natural selection could not have occurred. Astronomers became used to asking how finely balanced a situation existed in the Universe with respect to the existence of life like us or life of any other conceivable sort. The observed values of many of the fundamental constants of Nature or of the quantities describing the properties of the Universe's global properties – its shape, its speed of expansion, its uniformity – also seemed quite delicately poised. Quite small shifts in the *status quo* would render all conceivable complexity impossible. Habitable universes came to be seen as rather a tricky balancing act to accomplish.

LIFE IN AN EDWARDIAN UNIVERSE

'It is more important that a proposition be interesting than that it be true . . . But of course a true proposition is more apt to be interesting than a false one.'

Alfred North Whitehead[17]

It is interesting to end our look at Dicke's way of treating the Large Number coincidences between constants of Nature by taking a look back at a very similar type of argument made by Alfred Wallace in 1903. Wallace was a great scientist who today receives little of the credit he deserves. It was he, rather than Charles Darwin, who first had the idea that living organisms evolve by a process of natural selection. Fortunately for Darwin, who had been thinking deeply and gathering evidence to support such an idea independently of Wallace over a very long period of time, Wallace wrote to him to tell him of his ideas rather than simply publishing them in the scientific literature. Yet today, 'evolutionary biology' focuses almost entirely upon the contributions of Darwin.

Wallace was far broader in his interests than Darwin and was interested in most areas of physics, astronomy and earth sciences. In 1903 he published a wide-ranging study of the factors that make the Earth a habitable place and went on to explore the philosophical conclusions that might be drawn from the state of the Universe. His book went under the resonant title *Man's Place in the Universe*.[18] This was before the discovery of the theories of relativity, nuclear energy and the expanding universe.[19] Most nineteenth-century astronomers conceived of the Universe as a single island of matter, what we would now call our Milky Way galaxy. It was not established that there existed other galaxies or what the overall scale of the Universe was. It was only clear that it was big.

Wallace was impressed by the simple cosmological model that Lord Kelvin had developed using Newton's law of gravitation. It showed that if we took a very large ball of material then the action of gravity would cause it all to collapse towards its centre. The only way to avoid getting pulled into the centre was to orbit around the centre. Kelvin's universe contained about one billion stars like the Sun in order that their gravitational forces would counterbalance motions at the observed speeds.[20]

What is intriguing about Wallace's discussion[21] of this model of the Universe is that he adopts a non-Copernican attitude because he sees how some places in the Universe are more conducive to the presence of life than others. As a result, it is only to be expected that we lie near, but not at, the centre of things.

Remarkably, Wallace produces an analogue of Dicke's argument for the great age of any universe observed by humans. Of course, in Wallace's time, long before the discovery of nuclear power sources, no one knew how the Sun was powered. Kelvin had argued for gravitational energy, but it was not adequate for the job. In Kelvin's cosmology, material would be attracted by gravity into the central regions where the Milky Way was situated and fall into the stars that were already there, generating heat and maintaining their luminous power output over huge periods of time. Here, Wallace sees a simple reason for the vast size of the Universe:

'Here then, I think, we have found an adequate explanation of the very long-continued light- and heat-emitting capacity of our sun, and probably of many others in about the same position in the solar cluster. These would at first gradually aggregate into considerable masses from the slowly moving diffuse matter of the central portions of the original universe; but at a later period they would be reinforced by a constant and steady inrush of matter from its outer regions possessing such high velocities as to aid in producing and maintaining the requisite temperature of a sun such as ours, during the long periods demanded for continuous life-development. The enormous extension and mass of the original universe of diffused matter (as postulated by Lord Kelvin) is thus seen to be of the greatest importance as regards this ultimate product of evolution, because without it, the comparatively slow-moving and cool central regions might not have been able to produce and maintain the requisite energy in the form of heat; while the aggregation of by far the larger portion of its matter in the great revolving ring of the galaxy was equally important, in order to prevent the too great and too rapid inflow of matter to those favoured regions. . . . For [on] those [planets around stars] whose material evolution has gone on quicker or slower there has not been, or will not be, time enough for the development of life.'[22]

Wallace sees clearly the connection between these unusual global features of the Universe and the conditions necessary for life to evolve and prosper:

> 'we can dimly see the bearing of all the great features of the stellar universe upon the successful development of life. These are, its vast dimensions; the form it has acquired in the mighty ring of the Milky Way; and our position near to, but not exactly in, its centre.'[23]

He also expects that this process of infall and solar power generation from gravitational energy will probably have a staccato form with long periods of infall driving heating of the stars followed by periods of quiescence and cooling, a period which we have just begun to experience:

> 'I have here suggested a mode of development which would lead to a very slow but continuous growth of the more central suns; to an excessively long period of nearly stationary heat-giving power; and lastly, an equally long period of very gradual cooling – a period the commencement of which our sun may have just entered upon.[24]

Wallace completes his discussion of the cosmic conditions needed for the evolution of life by turning his attention to the geology and history of the Earth. Here he sees a far more complicated situation than exists in astronomy. He appreciates the host of historical accidents that have marked the evolutionary trail that has led to us, and thinks it 'in the highest degree improbable' that the whole collection of features that are conducive to the evolution of life will be found elsewhere. This leads him to speculate that the huge size of the Universe might be required in order to allow life a reasonable chance of developing on just one planet, like our own, no matter how conducive its local environment might be:

> 'such a vast and complex universe as that which we know exists around us, may have been absolutely required . . . in order to produce a world that should be precisely adapted in every detail for the orderly development of life culminating in man.'[25]

Today, we might echo this sentiment. The large size of the observable Universe, with its 10^{80} atoms, allows a vast number of sites for the statistical variations of chemical combination to be worked through.

Yet, despite his interest in the huge size of the Universe in making it probable that we evolved, Wallace was averse to the idea of a Universe populated with many other living beings. He believed that the uniformity of the laws of physics and chemistry[26] would ensure that

> 'organised living beings wherever they may exist in the universe must be fundamentally, and in essential nature, the same also. The outward forms of life, if they exist elsewhere, may vary, almost infinitely, as they do vary on earth . . . We do not say that organic life *could* not exist under altogether diverse conditions from those which we know or can conceive, conditions which may prevail in other universes constructed quite differently from ours, where other substances replace the matter and ether of our universe, and where other laws prevail. But *within* the universe we know, there is not the slightest reason to suppose organic life to be possible, except under the same general conditions and laws which prevail here.'[27]

Wallace provides an intriguing bridge between the pre-evolutionary way of thinking and the modern perspective brought about by the discovery that the Universe is changing. His approach to cosmology shows how the consideration of the conditions necessary for the evolution of life is not wedded to any particular theory of star formation and development but must be used in each context as appropriate. For Wallace it was a new picture of the Universe developed by Kelvin. For modern astronomers it is the well-tested theory of the expanding Universe in which the energy generation by the stars is almost completely understood. Both theories were dynamic: Kelvin's model allowed material to fall from great distances into the centre of the star system under the influence of gravitational attraction whilst the Big Bang theory of Dicke's expanded to increasing size with the passage of time. In both scenarios size and time were linked and the vastness of the Universe had unusual indirect consequences for what could happen within it, consequences that had a crucial bearing on the possibility of life and mind emerging in the course of time.

chapter eight

The Anthropic Principle

'Life is what the least of us make most of us feel the least of us make the most of.'

Willard Quine[1]

ANTHROPIC ARGUMENTS

'I have opinions of my own – strong opinions – but I don't always agree with them.'

President George W. Bush

Ever since these early realisations that there are properties of the Universe which are necessary for life, there has been growing interest in what has become known as the 'Anthropic Principle', and a wide-ranging debate has continued amongst astronomers, physicists and philosophers about its usefulness and ultimate significance. One of the reasons for this degree of interest has been the discovery that there are many ways in which the actual values of the constants of Nature help to make life a possibility in the Universe. Moreover, they sometimes appear to allow it to be possible by only a hair's breadth. We can easily imagine worlds in which the constants of Nature take on slightly different numerical values where living beings like ourselves would not be possible. Make the fine structure constant bigger and there can be no atoms, make the strength of gravity greater and stars exhaust their fuel very quickly, reduce the strength of nuclear forces and there can be no biochemistry, and so on. There are three types of change to consider. Tiny, infinitesimal, changes are possible. If we change the value of the fine structure constant only in the twentieth decimal place there will be no bad consequences for life that we know of. If we change it by a very small amount, say in the second decimal place, then the changes become more significant. Properties of atoms are altered and complicated processes like protein folding or DNA replication may be adversely affected. However, new possibilities for chemical complexity may open up. Evaluating the consequences of these changes is difficult because they are not so cut and dried. Third, there are very large changes. These will stop atoms or nuclei existing at all and are much more clear-cut as a barrier to developing complexity based upon the forces of Nature. For many conceivable changes, there could be no imaginable forms of life at all.

First, it is important to be quite clear about the way in which Dicke introduced his anthropic argument since there is considerable confusion[2] amongst commentators. A condition, like the existence of stars or certain chemical elements, is identified as a *necessary* condition for the existence of any form of chemical complexity, of which life is the most impressive known example. This does not mean that if this condition is met that life must exist, will never die out if it does exist, or that the fact that this condition holds in our Universe means that it was 'designed' with life in mind. These are all quite separate matters. If our 'necessary' anthropic condition is truly a necessary condition for living observers to exist in the Universe then it is a feature of the Universe that we must find it to possess, no matter how unlikely it might appear *a priori*.

The error that many people now make is to assume that an anthropic argument of this sort is a new scientific theory about the Universe that is a rival for other more conventional forms of explanation as to why the Universe possesses the 'necessary' anthropic condition. In fact, it is nothing of the sort. It is simply a methodological principle which, if ignored or missed, will lead us to draw incorrect conclusions. As we have seen, the story of Dirac and Dicke is a case in point. Dirac did not realise that a Large Number coincidence was a necessary consequence of being an observer who looks at the Universe at a time roughly equal to time required for stars to make the chemical elements needed for complex life to evolve spontaneously. As a result he drew the wrong conclusion that huge changes needed to be made to the laws of physics – changing the law of gravity to allow G to vary in time. Dicke showed that although such a coincidence might appear unlikely *a priori*, it was in fact a necessary feature of a universe containing observers like ourselves. It is therefore no more (and no less) surprising a feature of the Universe than our own existence.

There are many interesting examples of observer bias in less cosmic situations than the one considered by Dicke. My favourite concerns our perceptions of traffic flow. A recent survey of Canadian drivers[3] showed that they tended to think that the next lane on the highway moves faster than the one they are travelling in. This inspired the authors of the study to propose many complex psychological reasons for this belief amongst drivers, thinking perhaps that a driver is more likely to make comparisons with other traffic when being passed by faster cars than when overtaking them or that being overtaken leaves a bigger impression on a driver than overtaking. These conclusions are by no means unimportant because one of the conclusions of the study was that drivers might be educated about these tendencies so as to help them resist the constant urge to change lanes in search of a faster path, thereby speeding total traffic flow and improving safety. However, while the psychological causes might well be present, there is a simpler explanation for the results of the survey: *traffic does go faster in the other lanes!* The reason is a form of observer selection. Typically, slower lanes are created by overcrowding.[4] So, on the average, there are more cars in congested lanes moving slowly than there are in emptier lanes moving faster.[5] If you select a driver at random and ask them if they think the next lane is faster you are more likely to pick a driver in a congested lane because that's where most drivers are. Unfortunately, because of observer bias the driver survey does not tell

Figure 8.1 *Why do the cars in the other lane seem to be going faster? Because on the average they are!*[6]

you anything about whether it is good or bad to change lanes. The grass, maybe, is always greener on the other side.

Once we know of a feature of the Universe that is necessary for the existence of chemical complexity, it is often possible to show that other features of the Universe that appear to have nothing to do with life are necessary by-products of the 'necessary' condition. For example, Dicke's argument really tells us that the Universe has to be billions of years old in order that there be enough time for the building blocks of life to be manufactured in the stars. But the laws of gravitation tell us that the age of the Universe is directly linked to other properties it displays, like its density, its temperature, and the brightness of the sky. Since the Universe must expand for billions of years it must become billions of light years in visible extent. Since its temperature and density fall as it expands it necessarily becomes cold and sparse. As we have seen, the density of the Universe today is little more than one atom in every cubic metre of space. Translated into a measure of the average distances between stars or galaxies this very low density shows why it is not surprising that other star systems are so far away and contact with extraterrestrials is difficult. If other advanced life-forms exist in the Universe then, like us, they will have evolved unperturbed by beings from other worlds until they reached an advanced technological stage. Moreover, the very low temperature of radiation does more than ensure that space is a cold place. It guarantees the darkness of the night sky. For centuries scientists have wondered about this tantalising feature of the Universe. If there were huge numbers of stars out there in space then you might have thought that looking up into the night sky would be a bit like looking into a dense forest (Figure 8.2).

Figure 8.2 *If you look into a deep forest then your line of sight always ends on a tree.*[7]

Every line of sight should end on a star. Their shining surfaces would cover every part of the sky making it look like the surface of the Sun. What saves us from this shining sky is the expansion of the Universe. In order to meet the necessary condition for living complexity to exist there has to be ten billion years of expansion and cooling off. The density of matter has fallen to such a low value that even if all the matter were suddenly changed into radiant energy we would not notice any significant brightening of the night sky. There is just too little radiation to fill too great a space for the sky to appear bright any more. Once, when the universe was much younger, less than a hundred thousand years old, the whole sky was bright, so bright that neither stars nor atoms nor molecules could exist. Observers could not have been there to witness it.

These observations have other by-products of a much more philosophical nature. The large size and gloomy darkness of the Universe appear superficially to be deeply inhospitable to life. The appearance of the night sky is responsible for many religious and aesthetic longings born out of our apparent smallness and insignificance in the light of the grandeur and unchangeability of the distant stars. Many civilisations worshipped the stars or believed that they governed their future while others, like our own, often yearn to visit them.

George Santayana writes in *The Sense of Beauty*[8] of the emotional effect that results from the contemplation of the insignificance of the Earth and the vastness of the star-spangled heavens. For,

> 'The idea of the insignificance of our earth and of the incomprehensible multiplicity of worlds is indeed immensely impressive; it may even be intensely disagreeable . . . Our mathematical imagination is put on the rack by an attempted conception that has all the anguish of a nightmare and probably, could we but awake, all its laughable absurdity . . . the kinship of the emotion produced by the stars with the emotion proper to certain religious moments

makes the stars seem a religious object. They become, like impressive music, a stimulus to worship.

Nothing is objectively impressive; things are impressive only when they succeed in touching the sensibility of the observer, by finding the avenues to his heart and brain. The idea that the universe is a multitude of minute spheres circling like specks of dust, in a dark and boundless void, might leave us cold and indifferent, if not bored and depressed, were it not that we identify this hypothetical scheme with the visible splendour, the poignant intensity, and the baffling number of the stars.

. . . the sensuous contrast of the dark background, – blacker the clearer the night and the more stars we can see, – with the palpitating fire of the stars themselves, could not be exceeded by any possible device.'

Others have taken a more prosaic view. The heavyweight English mathematician and philosopher Frank Ramsey (the brother of Michael Ramsey, the former Archbishop of Canterbury) responded to Pascal's terror at 'the silence of the infinite spaces' around us in a sanguine fashion by remarking that,

'Where I seem to differ from some of my friends is in attaching little importance to physical size. I don't feel the least humble before the vastness of the heavens. The stars may be large, but they cannot think or love; and these are qualities which impress me far more than size does. I take no credit for weighing nearly seventeen stone. My picture of the world is drawn in perspective, and not like a model drawn to scale. The foreground is occupied by human beings, and the stars are all as small as threepenny bits.'[9]

Yet, although size isn't everything, on a cosmic scale it is most certainly something. The link between the time that the expansion of the Universe has apparently been going on for (which we usually call the 'age' of the Universe) and other things to do with life was something that cosmologists should have latched on to far more quickly. It could have stopped them pursuing another incorrect cosmological possibility for nearly twenty years. In 1948 Hermann Bondi, Thomas Gold and Fred Hoyle introduced a rival to the expanding Big Bang Universe. The Big Bang theory[10] implied that the expansion of the Universe began at a definite past time. Subsequently, the density and temperature of the matter and radiation in the Universe fell steadily as the Universe expanded. This expansion may continue forever or it may one day reverse into a state of contraction, revisiting conditions of ever greater density and temperature until a Big Crunch is encountered at a finite time in our future (see Figure 8.3).

This evolving scenario has the key feature that the physical conditions in the Universe's past were not the same as those that exist today or which will exist in the future. There were epochs when life could not exist because it was too hot for atoms to exist; there were epochs before there were stars and there will be a time when all the stars have died. In this scenario there is a preferred interval of cosmic history during which observers are most likely first to evolve and make their observations of the Universe. It also implied that there was a beginning to the Universe, a past time before which it (and time itself perhaps) did not exist, but it was silent as to the why or the wherefore of this beginning.

The alternative scenario created by Bondi, Gold and Hoyle was motivated in part by a desire to do away with the need for a beginning (or a possible end) to the Universe. Their other aim was to create a cosmological scenario that looked, on the average, always the same so that there were no preferred times in cosmic history (see Figure 8.4). At first this seems impossible to achieve. After all, the Universe *is* expanding. It is changing, so how can it be rendered changeless? Hoyle's vision was of a steadily flowing river, always moving but always much

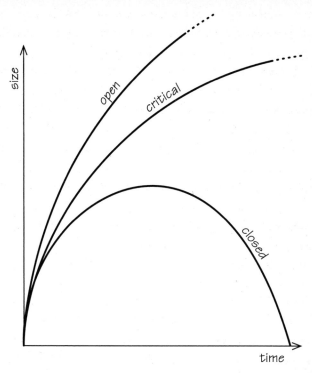

Figure 8.3 *The two types of expanding universe: 'open' universes expand forever; 'closed' universes eventually contract back to an apparent Big Crunch at a finite time in the future. The 'critical' universe marks the dividing line between the two and also expands for ever.*

the same. In order for the Universe to present the same average density of matter and rate of expansion no matter when it was observed, the density would need to be constant. Hoyle proposed that instead of the matter in the Universe coming into being at one past moment it was continuously created at a rate that exactly countered the tendency for the density to be diluted by the expansion. This mechanism of 'continuous creation' needed to occur only very slowly to achieve a constant density; only about one atom in every cubic metre every 10 billion years was required and no experiment or astronomical observation would

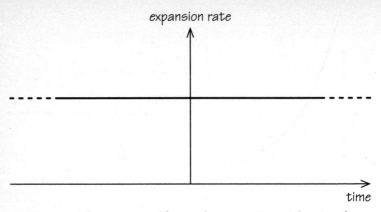

Figure 8.4 *The expansion of a steady-state universe. The rate of expansion is always the same. There is no beginning and no end, no special epoch when life can first emerge or after which it begins to die out along with the stars. The universe looks the same on the average at all times in its history.*

be able to detect an effect so small. This 'steady state' theory of the Universe made very definite predictions. The Universe looked the same on the average at all times. There were no special epochs in cosmic history – no 'beginning', no 'end', no time when stars started to form or when life first became possible in the Universe (see Figure 8.5).

Eventually this theory was ruled out by a sequence of observations that began in the mid 1950s, showing first that the population of galaxies that were profuse emitters of radio waves varied significantly as the Universe aged, and culminated in the discovery in 1965 of the heat radiation left over from the hot beginning predicted by the Big Bang models. This microwave background radiation had no place in the steady state Universe.

For twenty years astronomers tried to find evidence that would tell us whether the Universe was truly in the steady state that Bondi, Gold, and Hoyle proposed. A simple anthropic argument could have shown how unlikely such a state of affairs would be. If you measure the expansion rate of the Universe it tells you a time for which the Universe

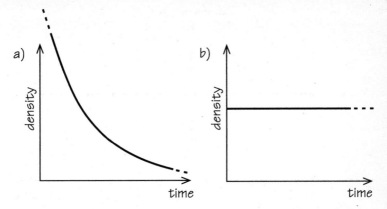

Figure 8.5 *(a) The variation of the average density of matter in an expanding Big Bang universe. (b) The average density of matter in a steady-state universe is always the same.*

appears to have been expanding.[11] In a Big Bang universe this really is the time since the expansion began – the age of the Universe. In the steady state theory there is no beginning and the expansion rate is just the expansion rate and nothing more. This is illustrated in the picture shown in Figure 8.4.

In a Big Bang theory the fact that the expansion age is just slightly greater than the age of the stars is a natural state of affairs. The stars formed in our past and so we should expect to find ourselves on the cosmic scene after they have formed. But in a steady state universe the 'age' is infinite and is not linked to the rate of expansion. In a steady state Universe it is therefore a complete coincidence that the inverse of the expansion rate gives a time that is roughly equal to the time required for stars to produce elements like carbon. Just as surely as the coincidence between the inverse of the expansion rate of the Universe and the time required for stars to produce biochemical elements ruled out the need for Dirac's varying G, it should have cast doubt on the need for a steady state universe.

A DELICATE BALANCE

'A banker is a man who lends you an umbrella when the
weather is fair, and takes it away from you when it rains.'

Mark Twain

We have seen that a good deal of time is needed if stars are to manu-
facture carbon out of inert gases like hydrogen and helium. But time
is not enough. The specific nuclear reaction that is needed to make
carbon is a rather improbable one. It requires three nuclei of helium
to come together to fuse into a single nucleus of carbon. Helium nuclei
are called alpha particles and this key carbon-forming reaction has been
dubbed the 'triple-alpha' process. The American physicist Ed Salpeter
first recognised its significance in 1952. However, a few months later,
whilst visiting Cal Tech in Pasadena, Fred Hoyle realised that making
carbon in stars by this process was doubly difficult. First, it was difficult to get three alpha particles to meet and, even if you did, the fruits
of their liaison might be short-lived. For if you looked a little further
down the chain of nuclear reactions it seemed that all the carbon could
quickly get consumed by interacting with another alpha particle to
create oxygen.

Hoyle realised that the only way to explain why there was a significant amount of carbon in the universe was if the production of carbon
went much faster and more efficiently than had been envisaged, so that
the ensuing burning to oxygen did not have time to destroy it all. There
was only one way to achieve this carbon boost. Nuclear reactions occasionally experience special situations where their rates are dramatically
increased. They are said to be 'resonant' if the sum of the energies of
the incoming reacting particles is very close to a natural energy level
of a new heavier nucleus. When this happens the nuclear reaction goes
especially quickly, often by a huge factor.

Hoyle saw that the presence of a significant amount of carbon in the Universe would be possible only if the carbon nucleus possessed a natural energy level at about 7.65 MeV above the ground level. Only if that was the case could the cosmic carbon abundance be explained, Hoyle reasoned. Unfortunately no energy level was known in the carbon nucleus at the required place.[12]

Pasadena was a good place to think about the energy levels of nuclei. Willy Fowler led a team of outstanding nuclear physicists and was an extremely affable and enthusiastic individual. Hoyle didn't hesitate to pay him a visit. And soon Fowler had persuaded himself that all the past experiments could indeed have missed the energy level that Hoyle was proposing. Within days, Fowler had pulled in other nuclear physicists from the Kellogg Radiation Lab and an experiment was planned. The result when it came was dramatic.[13] There was a new energy level in the carbon nucleus at 7.656 MeV, just where Hoyle had predicted it would be.

The whole sequence of events for the production of carbon by stars then looked so delicately balanced that, as a science-fiction universe, it would have seemed contrived. First, three helium nuclei (alpha particles) have to interact at one place. This they manage to do in a two-step process. First, two helium nuclei combine to create a beryllium nucleus

$$\text{helium} + \text{helium} \rightarrow \text{beryllium}$$

Fortunately, beryllium has a peculiarly long lifetime,[14] ten thousand times longer than the time required for two helium nuclei to interact and so it stays around long enough to have a good chance of combining with another helium nucleus to produce a carbon nucleus:

$$\text{beryllium} + \text{helium} \rightarrow \text{carbon}$$

The 7.656 MeV energy level in the carbon nucleus lies *just* above the energies of the beryllium plus helium (7.3667 MeV), so that when the thermal energy of the inside of the star is added the nuclear reaction becomes resonant and lots of carbon is produced. But that is not the end of the story. The next reaction waiting to burn up all the carbon is

$$\text{carbon} + \text{helium} \rightarrow \text{oxygen.}$$

What if this reaction should turn out to be resonant as well? Then all the rapidly produced carbon would disappear and the carbon resonance level would be to no avail. Remarkably, this last reaction just fails to be resonant. The oxygen nucleus has an energy level at 7.1187 MeV that lies just *below* the total energy of carbon plus helium at 7.1616 MeV. So when the extra thermal energy in the star is added this reaction can never be resonant and the carbon survives (see Figure 8.6). Hoyle recognised that his finely balanced sequence of apparent co-incidences was what made carbon-based life a possibility in the Universe.[15]

The positioning of the nuclear energy levels in carbon and oxygen is the result of a very complicated interaction between nuclear and electromagnetic forces that could not be calculated easily when the discovery of the carbon resonance level was first made. Today, it is possible to make very good estimates of the contributions of electromagnetic and nuclear forces to the levels concerned. One can see that their positions are a consequence of the fine structure constant and strong nuclear force constant taking the values that they do with high precision. If the fine structure constant, that governs the strength of electromagnetic forces, were changed by more than 4 per cent or the strong force by more than 0.4 of one per cent then the production of carbon or oxygen would be reduced by factors of between 30 and 1000. More detailed calculations of the fate of stars when these constants of Nature are slightly changed have been carried out recently by Heinz Oberhummer, Attila Csótó and Helmut Schlattl.[16] Their results can be seen in Figure 8.6.

We see that the carbon and oxygen levels vary systematically as the constants of Nature governing the position of the resonance levels

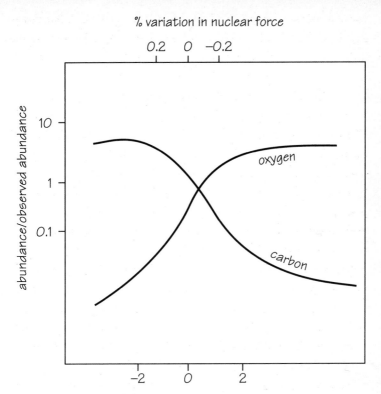

Figure 8.6 *The production of carbon and oxygen by stars if the constants of Nature governing the strengths of the electromagnetic and nuclear forces are changed by the indicated amounts.*

are changed. If they are altered from their actual values we end up with large amounts of carbon or large amounts of oxygen but never of both. A change of more than 0.4 per cent in the constants governing the strength of the strong nuclear force or more than 4 per cent in the fine structure constant would destroy almost all carbon or almost all oxygen in every star.

Hoyle had been very struck by the carbon resonance-level coincidence and its implications for the constants of physics. Rounding off a discussion of the astrophysical origin of the elements, he wrote,[17]

> 'But I think one must have a modicum of curiosity about the strange dimensionless numbers [constants] that appear in physics, and on which, in the last analysis, the precise positioning of the levels in a nucleus such as C^{12} or O^{16} must depend. Are these numbers immutable, like the atoms of the nineteenth-century physicist? Could there be a consistent physics with different values for the numbers?'

Hoyle sees two alternatives on offer: either we must seek to demonstrate that the actual values of the constants of Nature 'are all entirely necessary to the logical consistency of physics', or adopt the point of view that 'some, if not all, the numbers in question are fluctuations; that in other places of the Universe their values would be different.'

At first, Hoyle favoured the second 'fluctuation' idea, that the constants of Nature might be varying, possibly randomly, throughout space so that only in some places would the balance between the fine structure constant and the strong force constant come out 'just right' to allow an abundance of carbon and oxygen. Thus, if this picture is adopted,[18]

> 'the curious placing of the levels in C^{12} and O^{16} need no longer have the appearance of astonishing accidents. It could simply be that since creatures like ourselves depend on a balance between carbon and oxygen, we can exist only in the portions of the universe where these levels happen to be correctly placed. In other places the level on O^{16} might be a little higher, so that the addition of α-particles to C^{12} was highly resonant. In such a place . . . creatures like ourselves could not exist.'

In the years that followed Hoyle gradually took a more deterministic view of the resonance level coincidences, seeing them as evidence for some form of pre-planning of the universe in order to make life possible:[19]

> 'I do not believe that any scientist who examined the evidence would fail to draw the inference that the laws of nuclear physics have been deliberately designed with regard to the consequences they produce inside the stars. If this is so, then my apparently random quirks have become part of a deep-laid scheme. If not then we are back again at a monstrous sequence of accidents.'

Hoyle's successful prediction sparked a resurgence of interest in the old Design Arguments, beloved of eighteenth- and nineteenth-century natural theologians, but with a new twist. Since ancient times there had been strong support for an argument for the existence of God (or 'the gods') from the fact that living things seemed to be tailor-made for their function. Animals appeared to be perfectly camouflaged for their environments; parts of our bodies were delicately engineered to provide (most of) us with ease of mobility, good vision, keen hearing and so forth;[20] the motions of the planets were favourably arranged to make the terrestrial climate conducive to the continuance of life. Large numbers of apparent coincidences existed between things and persuaded many a philosopher, theologian and scientist of the past that none of this was an accident. The Universe was designed with an end in view. This end involved the existence of life – perhaps even ourselves – and the plainness of the evidence for such design meant that there had to be a Designer.

As it stands this ancient argument was difficult to refute by means of scientific facts. And it was always persuasive for those who were not scientists. After all, there *are* remarkable adaptations between living things and their environments all over the natural world. It is much easier to undermine by means of logical or philosophical argument. But scientists are never very impressed by such arguments unless they can provide a better explanation. And so it was with the Design Argument. Despite its blinkered attitude to many of the realities of the world, it was only overthrown as a serious explanation of the existence of complexity in Nature when a better explanation came along.[21] The better explanation was by means of evolution by natural selection, which showed how living things can become well adapted to their environments over the course of time under a very wide range of circumstances, so long as the environment is not changing too quickly. Complexity could develop from simplicity without direct Divine intervention.

It is important to see what this type of Design Argument was focusing on. It is an argument about the inter-relationships between different *outcomes* of the laws of Nature. These are only partially determined by the forms of the laws. Their form is also determined by the constants of Nature, the starting conditions, and all manner of other statistical accidents.[22]

In the late seventeenth century Isaac Newton discovered the laws of motion, gravitation and optics that enabled us to understand the workings of the inanimate world around us and the motions of the celestial bodies in remarkable detail. Newton's success was seized upon by natural theologians and religious apologists who saw the beginnings of another style of Design Argument altogether – one based not upon the outcomes of the laws of Nature but upon the form of the laws themselves. With Newton's encouragement there grew up a Design Argument based upon the evident intelligence, mathematical elegance and effectiveness of Newton's *laws* of Nature. A typical form of the argument would be to show that the famous inverse-square law of gravitation was optimal for the existence of a solar system. If it had been an inverse-cube or any other inverse power of distance other than two then there could not exist stable planetary periodic orbits. All planets would follow a spiral path into the Sun or escape to infinity. This type of argument is quite different from the teleological form based upon fortuitous outcomes and adaptations. It identifies the most deep-seated basis for 'order' in the Universe as the fact that it can be so widely and accurately described by simple mathematical laws. It then presumes that order needs an 'orderer'.

The contrast between these two forms of the Design Argument – from laws and from outcomes – is clearly displayed by the effects of the discovery that organisms evolve by natural selection. This quickly finished the argument from outcomes as a useful explanation of anything.[23] But the Design Argument based on laws was completely unaffected. Natural selection did not act upon laws of motion or forces of Nature nor, as Maxwell liked to stress, could selection alter the properties of atoms and molecules.

In retrospect, it is clear that it is possible to create a further, distinct form of the Design Argument which appeals to the particular values taken by the fundamental constants of Nature. It is this set of numbers that distinguishes our Universe from others, and fixes the resonance levels in carbon and oxygen nuclei. It would be possible for the laws of Nature that we know to take the same form yet for the constants of Nature to change their values. The outcomes would then be very different.

The fact that we can shift the values of constants of Nature in so many of our laws of Nature may be a reflection of our ignorance. Many physicists believe, like Eddington, that ultimately the values of the constants of Nature will be shown to be inevitable and we will be able to calculate them in terms of pure numbers. However, it has become increasingly clear, as we will see in later chapters, that not all of the constants will be determined in this way. Moreover, the nature of the determination for the others may have a significant statistical aspect. What may be predicted is not *the* value but a probability distribution that the constant take any value. There will no doubt be a most probable value but that may not be the value that we see, if only because it may characterise a universe in which observers cannot exist.

BRANDON CARTER'S PRINCIPLES

'I do not feel like an alien in this universe. The more I examine the universe and study the details of its architecture, the more evidence I find that the universe in some sense must have known that we were coming.'

Freeman Dyson[24]

The general importance of Dicke's approach to understanding the Large Numbers of cosmology was first seized upon by Brandon Carter, then a Cambridge astrophysicist but now working at Meudon in Paris. Carter had learned about the Large Number coincidences by reading Bondi's student textbook of cosmology[25] but had not fallen under the spell of the steady state theory that was the centrepiece of Bondi's presentation. Bondi liked to assume that because the laws of Nature must be always the same, all other gross aspects of the Universe should display the same uniformity in space and time.[26] The steady state theory was based on precisely this premise – that the gross structure of the Universe is always the same on the average. Bondi confessed to not having been able to follow the calculations of Eddington in his attempts to explain the Large Numbers by means of his Fundamental Theory. By contrast, he is more outspoken about Dirac's scheme to render the gravitation constant a time variable, seeing it as a further denial of the steady state principle:

'Dirac . . . is contraposed to the basic arguments of the steady-state theory, since it supposes that not only the universe changes but with it the constants of atomic physics. In some ways it may almost be said to strengthen the steady-state arguments by showing how limitless the variations are that may be imagined to arise in a changing universe.'[27]

As a result of considering Dicke's explanation for the inevitability of our observation of some of the Large Number coincidences, Carter saw that it was important to stress the limitations of grand philosophical assumptions about the uniformity of the Universe. Ever since Copernicus showed that the Earth should not be placed at the centre of the known astronomical world, astronomers had used the term Copernican Principle to underwrite the assumption that we must not assume anything special about our position in the Universe. Einstein had assumed this implicitly when he first searched for mathematical descriptions of the Universe by seeking solutions of his equations that ensured every place in the Universe was the same: same density, same rate of expansion, and same temperature. The steady-statesmen went one step further by seeking universes that were the same *at every time* in cosmic history as well. Of course, the real Universe cannot be *exactly* the same everywhere but to a very good approximation, when one averages over large enough regions of space, it appears to be so, to an accuracy of about one part in one hundred thousand.

Carter rejected the wholesale use of the Copernican Principle in more specific situations because there are clearly restrictions on where and when observers could be present in the Universe:

> 'Copernicus taught us the very sound lesson that we must not assume gratuitously that we occupy a privileged *central* position in the Universe. Unfortunately there has been a strong (not always subconscious) tendency to extend this to a questionable dogma to the effect that our situation cannot be privileged in any sense.'[28]

Carter's emphasis upon the role of the Copernican Principle was encouraged by the fact that this presentation took the form of a lecture at an international astronomy meeting convened in Cracow to coincide with the 500th anniversary of Copernicus' birth.

Dicke's argument showed that there was good reason to expect life to come on the scene several billion years after the expansion of a Big Bang Universe began. This showed that one of the Large Numbers coincidences was an inevitable observation by such observers. This was an application of what Carter called the *weak anthropic principle*,

> 'that what we can expect to observe must be restricted by the condition necessary for our presence as observers.'[29]

Later, Carter regretted using the term 'anthropic principle'. The adjective 'anthropic' has been the source of much confusion because it implies there is something in this argument that focuses upon *Homo sapiens*. This is clearly not the case. It applies to all observers regardless of their form and biochemistry. But if they were not biochemically constructed from the elements that are made in the stars then the specific feature of the Universe that would be inevitable for them might differ from what is inevitable for us. However, the argument is not really changed if beings are possibly based upon silicon chemistry or physics. All the elements heavier than the chemically inert gases of hydrogen, deuterium and helium are made in the stars like carbon and require billions of years to create and distribute. Later, Carter preferred the term 'self-selection principle' to stress the way in which the necessary conditions for the existence of observers select, out of all the possible universes, some subset which allows observers to exist. If you are unaware that being an observer in the Universe already limits the type of universe you could expect to observe then you are liable to introduce unnecessary grand principles or unneeded changes to the laws of physics to explain unusual aspects of the Universe. The archetypal examples are Gerald Whitrow's discussion of the age and density of the universe[30] and Robert Dicke's explanation of the Large Numbers.

Carter's consideration of the self-selecting influence of our existence upon the sort of astronomical observations we make was inspired by reading about the Large Numbers coincidences in Bondi's book. Not knowing of Dicke's arguments of 1957 and 1961, he also noticed the importance of considering the inevitability of our observing the Universe close to the typical hydrogen-burning lifetime of a typical star. He was struck by the unnecessary way that Dirac had introduced the hypothesis of varying constants to explain these coincidences:[31]

> 'it was completely erroneous for him to have used this coincidence as a motivation for a radical departure from standard theory.
>
> At the time when I originally noticed Dirac's error, I simply supposed that it had been due to an emotionally neutral oversight, easily explicable as being due to the rudimentary state of general understanding of stellar evolution in the pioneering era of the 1930s, and that it was therefore likely to have been already recognised and corrected by its author. My motivation in bothering to formulate something that was (as I thought) so obvious as the anthropic principle in the form of an explicit precept, was partly provided by my later realisation that the source of such (patent) errors as that of Dirac was not limited to chance oversight or lack of information, but that it was also rooted in more deep seated emotional bias comparable with that responsible for early resistance to Darwinian ideas at the time of the "apes or angels" debates in the last century. I became aware of this in Dirac's own case when I learned of his reaction when his attention was explicitly drawn to the "anthropic" line of reasoning [about the Large Number coincidences] . . . when it was first pointed out by Dicke in 1961. This reaction amounted to a straight refusal to accept the line of reasoning leading to Dicke's (in my opinion unassailable) conclusion that "the statistical support for Dirac's cosmology is found to be missing". The reason offered by Dirac is rather astonishing in the context of a modern scientific debate: after making an unsubstantiated

(and superficially implausible) claim to the effect that in his own theory "life need never end" his argument is summarised by the amazing statement that, in choosing between his own theory and the usual one . . . "I prefer the one that allows the possibility of endless life". What I found astonishing here was of course the suggestion that such a preference could be relevant in such an argument . . . Dirac's error provides a salutary warning that contributes to the motivation for careful formulation of the anthropic and other related principles.'[32]

The weak anthropic principle applies naturally to help us understand why variable quantities take the range of values that we find them to take in our vicinity of space and time. But there exist 'coincidences' between combinations of quantities which are believed to be true constants of Nature. We will not be able to explain these coincidences by the fact that we live when the Universe is several billion years old, in conditions of relatively low density and temperature. Carter's suggested response to this was more speculative. If the constants of Nature can't change and are programmed into the overall structure of the Universe in a unique way then maybe there is some as yet unknown reason why there have to be observers in the Universe at some stage in its history? Carter dubbed this the *strong anthropic principle*, which states

'that the Universe (and hence the fundamental parameters on which it depends) must be such as to admit the creation of observers within it at some stage.'

The introduction of such a speculation needs evidence to support it. In this case it is that there are a number of unusual apparent coincidences between superficially unrelated constants of Nature that appear to be crucial for the existence of ourselves or any other conceivable form of life. Hoyle's unusual carbon and oxygen resonance levels are archetypal examples. There are many others. Small changes in the strengths of the different forces of Nature and in the masses of different elementary particles destroy many of the delicate balances that make life possible. By contrast, if the conditions for life to develop and persist had been found to depend only very weakly on all the constants of Nature then there would be no motivation for thinking about an anthropic principle of this stronger sort. In future chapters we shall see how this idea provokes a serious consideration of the idea that there exist other 'universes' which possess different properties and different constants of Nature so that we might conclude that we find ourselves inhabiting one of the possible universes in which the constants and cosmic conditions have fallen out in a pattern that permits life to exist and persist — for we could not find it otherwise.

A CLOSE-RUN THING?

'Do I dare
Disturb the universe?'

T.S. Eliot[33]

We have been saying that the values of the constants of Nature are rather fortuitously 'chosen' when it comes to allowing life to evolve and persist. Let's take a look at a few more examples. The structure of atoms and molecules is controlled almost completely by two numbers that we encountered in Chapter 5: the ratio of the electron and proton masses, β, which is approximately equal to $1/1836$ and the fine structure constant α, which is approximately equal to $1/137$. Suppose that we allow these two constants to change their values independently and we also assume (for simplicity) that no other constants of Nature are changed. What happens to the world if the laws of Nature stay the same?

If we follow up the consequences we soon find there isn't much room to manoeuvre. Increase β too much and there can be no ordered molecular structures because it's the small value of β that ensures that electrons occupy well-defined positions around an atomic nucleus and don't wiggle around too much. If they did then very fine-tuned processes like DNA replication would fail. The number β also plays a role in the energy generation processes that fuel the stars. Here it links up with α to make the centres of stars hot enough to initiate nuclear reactions. If β exceeds about $0.005 \, \alpha^2$ then there would be no stars. If modern grand unified gauge theories are on the right track then alpha must lie in the narrow range between about $1/180$ and $1/85$ otherwise protons will decay long before the stars can form. The Carter condition is also shown dashed $(---)$ on the picture. Its track picks out worlds where stars have convective outer regions which seem to be needed to make some systems of planets. The regions of α and β that are allowed and forbidden are shown in Figure 8.7.

If, instead of α *versus* β, we play the game of changing the strength of the strong nuclear force, α_s, together with that of α, then unless $\alpha_s > 0.3\alpha^{1/2}$ the biologically vital elements like carbon would not exist and there would not be any organic chemists. They would be unable to hold themselves together. If we increase α_s by just 4 per cent there is a potential disaster because a new nucleus, helium-2 made of two protons and no neutrons, can now exist[34] and allows very fast direct nuclear reactions proton + proton → helium-2. Stars would rapidly exhaust their fuel and collapse to degenerate states or black holes. In contrast, if α_s were decreased by about 10 per cent then the deuterium nucleus would cease to be bound and the nuclear astrophysical pathways to the biological elements would be blocked. Again, we find a rather small region of parameter space in which the basic building blocks of chemical complexity can exist. The inhabitable window is shown in Figure 8.8.

The more simultaneous variations of other constants one includes in these considerations, the more restrictive is the region where life, as

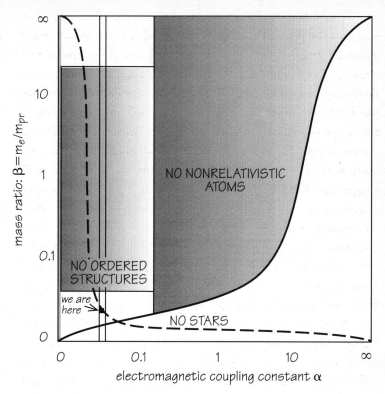

Figure 8.7 *The habitable zone where life-supporting complexity can exist if the values of β and α were permitted to vary independently. In the lower right-hand zone there can be no stars. In the upper right-hand zone there are no non-relativistic atoms. In the top left zone the electrons are insufficiently localised for there to exist highly ordered self-reproducing molecules. The narrow 'tramlines' pick out the region which may be necessary for matter to remain stable for long enough for stars and life to evolve.*[35]

we know it, can exist. It is very likely that if variations can be made then they are not all independent. Rather, making a small change in one constant might alter one or more of the others as well. This would tend to make the restrictions on most variations become even more tightly constrained.

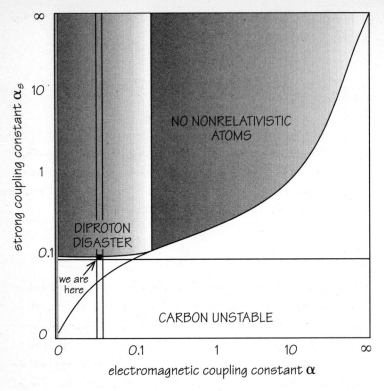

Figure 8.8 *The habitable zone where life-supporting complexity can exist if the values of* α_s *and* α *are changed independently. The bottom right-hand zone does not allow essential biochemical elements like carbon, nitrogen and oxygen to exist. The top left-hand zone permits a new nucleus, helium-2, called the diproton, to exist. This provides a route for very fast hydrogen burning in stars and would probably lead them to exhaust their fuel long before conditions were conducive to the formation of planets or the biological evolution of complexity.*[36]

These examples should be regarded as merely indications that the values of the constants of Nature are rather bio-friendly. If they are changed by even a small amount the world becomes lifeless and barren instead of a home for interesting complexity. It was this unusual state of affairs that first provoked Brandon Carter to see what sort of 'strong anthropic' explanations might be offered for the values of the constants of Nature.

SOME OTHER ANTHROPIC PRINCIPLES

'I don't want to achieve immortality through my work. I want to achieve immortality through not dying. I don't want to live on in the hearts of my countrymen. I would rather live on in my apartment.'

Woody Allen[37]

Other more speculative anthropic principles have been suggested by other researchers. John Wheeler, the Princeton scientist who coined the term 'black hole' and played a major role in their investigation, proposed what he called the *Participatory Anthropic Principle*. This is not especially to do with constants of Nature but is motivated by the fineness of the coincidences that allow life to exist in the cosmos. Perhaps, Wheeler asks, life is in some way essential for the coherence of the Universe? But surely *we* are of no consequence to the far-flung galaxies and the existence of the Universe in the distant past before life could exist? Wheeler was tempted to ask if the importance of observers in bring-ing quantum reality into full existence may be trying to tell us that 'observers', suitably defined, may be in some sense necessary to bring the Universe into existence. This is very hard to make good sense of because in quantum theory the notion of an observer is not sharply defined. It is anything that registers information. A photographic plate would do just as well as a night watchman.

A fourth Anthropic Principle, introduced by Frank Tipler and myself, is somewhat different. It is just a hypothesis that should be able to be shown to be true or false using the laws of physics and the observed state of the Universe. It is called the *Final Anthropic Principle* (or conjecture) and proposes that once life emerges in the Universe it will not die out. Once we have come up with a suitably wide definition of life, say as information processing ('thinking') with the ability to store information ('memory'), we can investigate whether this could be true. Note that there is no claim that life has to arise or that it must endure. Clearly, if life is to endure forever it must ultimately change its basis from life as we know it. Our knowledge of astrophysics tells us that the Sun will eventually undergo an irreversible energy crisis, expand, and engulf the Earth and the rest of the inner solar system. We will need to be gone from Earth by then, or to have transmitted the information needed to recreate members of our species (if it can still so be called) elsewhere. Thinking millions of years to the future we might also imagine that life will exist in forms that today would be called 'artificial'. Such forms might be little more than processors of information with a capacity to store information for future use. Like all forms of life they will be subject to evolution by natural selection.[38] Most likely they will be tiny. Already we see a trend in our own technological societies towards the fabrication of smaller and smaller machines that consume less and less energy and produce almost no waste. Taken to its logical conclusion, we expect advanced life-forms to be as small as the laws of physics allow.

In passing we might mention that this could explain why there is no evidence of extraterrestrial life in the Universe. If it is truly advanced, even by our standards, it will most likely be very small, down on the molecular scale. All sorts of advantages then accrue. There is lots of room there – huge populations can be sustained. Powerful, intrinsically quantum computation can be harnessed. Little raw material is required and space travel is easier. You can also avoid being detected by civilisations of clumsy bipeds living on bright planets that beam continuous radio noise into interplanetary space.

We can now ask whether the Universe allows information processing to continue forever. Even if you don't want to equate information processing with life, however futuristic, it should certainly be necessary for it to exist. This turns out to be a question that we can go quite close to answering. If the Universe began to accelerate a few billion years ago, as recent observations indicate, then it is likely that it will continue accelerating forever.[39] It will never slow down and contract back to a Big Crunch. If so, then we learn that information processing will come to a halt. Only a finite number of bits of information can be processed in a never-ending future. This is bad news. It occurs because the expansion is so rapid that information quality is very rapidly degraded.[40] Worse still, the accelerated expansion is so fast that light signals sent out by any civilisation will have a horizon beyond which they cannot be seen. The Universe will become partitioned into limited regions within which communication is possible.

An interesting observation was made along with the original proposal of the Final Anthropic Principle. We pointed out[41] that if the expansion of the Universe were found to be accelerating then information processing must eventually die out. Recently, important observational evidence has been gathered by several research groups to show that the expansion of the Universe began to accelerate just a few billion years ago. But suppose the observational evidence for the present acceleration of the Universe turns out to be incorrect.[42] What then? It is most likely that the Universe will keep on expanding forever but continuously decelerate as it does so. Life still faces an uphill battle to survive indefinitely. It needs to find differences in temperature, or density, or expansion in the Universe from which it can extract useful energy by making them uniform. If it relies on mining sources of energy that exist locally – dead stars, evaporating black holes, decaying elementary particles – then eventually it runs into the problem that well-worked coal mines inevitably face: it costs more to extract the energy than can be gained from it. Beings of the far future will find that they need to economise on energy usage – economise on living in fact! They can reduce their free energy consumption by spending long periods hibernating, waking up to process information for a while before returning to their inactive state. There is one potential problem with this Rip van Winkle existence.

You need a wake-up call. Some physical process needs to be arranged which will supply an unmissable wake-up call without using so much energy that the whole point of the hibernation period is lost. So far it is not clear whether this can be done forever. Eventually it appears that mining energy gradients that can be used to drive information processing becomes cost ineffective. Life must then begin to die out.

By contrast, if life does not confine its attentions to mining local sources of energy the long-range forecast looks much brighter. The Universe does not expand at exactly the same rate in every direction. There are small differences in speed from one direction to another which are attributable to gravitational waves of very long, probably infinite, wavelength threading space. The challenge for super-advanced life-forms is to find a way of tapping into this potentially unlimited energy supply. The remarkable thing about it is that its density falls off far more slowly than that of all ordinary forms of matter as the Universe expands. By exploiting the temperature differences created by radiation moving parallel to direction of expansion moving at different rates, life could find a way to keep its information processing going.

Lastly, if the Universe does collapse back to a future Big Crunch in a finite time then the prospects at first seem hopeless. Eventually, the collapsing Universe will contract sufficiently for galaxies and stars to merge. Temperatures will grow so high that molecules and atoms will be dismembered. Again, just as in the far future, life has to exist in some abstract disembodied form, perhaps woven into the fabric of space and time. Amazingly, it turns out that its indefinite survival is not ruled out so long as time is suitably defined. If the true time on which the universe 'ticks' is a time created by the expansion itself then it is possible for an infinite number of 'ticks' of this clock to occur in the finite amount of time that appears to be available on our clocks before the Big Crunch is reached.

There is one last trick that super-advanced survivors might have up their sleeves in universes that seem doomed to expand forever. In 1949 the logician Kurt Gödel, Einstein's friend and colleague at

Princeton, shocked him by showing that time travel was allowed by Einstein's theory of gravity.[43] He even found a solution of Einstein's equations for a universe in which this occurred. Unfortunately, Gödel's universe is nothing like the one that we live in. It spins very rapidly and disagrees with just about all astronomical observations one cares to make. However, there may be other more complicated possibilities that resemble our Universe in all needed respects but which still permit time travel to occur. Physicists have spent quite a lot of effort exploring how it might be possible to create the distortions of space and time needed for time travel to occur. If it is possible to engineer the conditions needed to send information backwards in time then this offers a strategy for escape from a lifeless future for suitably ethereal forms of 'life' defined by information processing and storage. Don't invest your efforts in perfecting means of extracting usable energy from an environment that is being driven closer and closer to a lifeless equilibrium. Instead, travel backwards in time to an era where conditions are far more hospitable. Indeed, travel is not strictly necessary, just transmit the instructions needed for re-emergence.

Often, people are worried about apparent factual paradoxes that can emerge from allowing backward time travel. Can't you kill yourself or your parents in infancy so that you cannot exist? All these paradoxes are impossibilities. They arise because you are introducing a physical and logical impossibility by hand. It helps to think of space and time in the way that Einstein taught us: as a single block of spacetime, see Figure 8.9.

Now step outside spacetime and look in at what happens there. Histories of individuals are paths through the block. If they curve back upon themselves to form closed loops then we would judge time travel to occur. But the paths are what they are. There is no history that is 'changed' by doing that. Time travel allows us to be part of the past but not to change the past. The only time-travelling histories that are possible are self-consistent paths. On any closed path there is no well-defined division between the future and the past. It is like having a

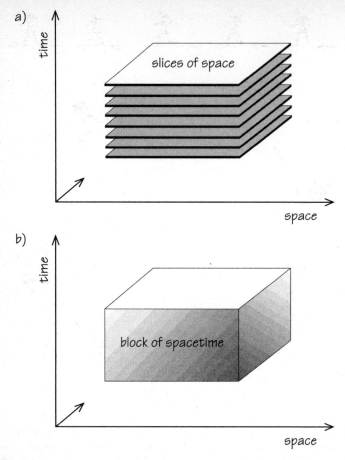

Figure 8.9 *(a) A stack of slices of space taken at different times; (b) a block of spacetime made from all the slices of space. This block could be sliced up in many ways that differ from the slicing chosen in (a).*

troop of soldiers marching one behind the other in single file. If they march in a straight line then it is clear who is in front of whom. But make them march in a circle so the previous leader follows the previous back-marker and there is no longer any well-defined sense of order in the line, as pictured in Figure 8.10.

Figure 8.10 *March in a straight line and it is clear who is in front of whom. March in a circle and everyone is in front and behind everyone else.*

If this type of backward time travel is an escape from the thermodynamic end of the Universe and our Universe appears to be heading for just such a thermodynamic erasure of all possibilities for processing information then maybe super-advanced beings in our future are already travelling backwards into the benign cosmic environment that the present-day universe affords. Many arguments have been put forward to argue against the arrival of tourists from the future but they have a rather anthropocentric purpose in mind. It has been argued that the great events in Earth's history (events around Bethlehem in 4 BC, the Crucifixion, the death of Socrates, and so on) would become magnets for backward time travellers, creating a huge cumulative audience that was evidently not present. But there is no reason why escapees from the heat death of the Universe should visit *us*, let alone cause crowd control problems at critical points in our history.

My favourite argument[44] against backward time travel is a financial one. It appeals to the fact that interest rates in the money markets are non-zero to argue that neither forward nor backward time travellers are taking advantage of their position to make a killing on the financial markets. If they were able to invest in the past on the basis of knowing where markets would increase in the future then the long-term result would be to drive interest rates to zero. Again, it is easy to avoid allowing this argument to rule out time travellers escaping the heat death of the Universe. One suspects, however, that financial investments might be the least of their concerns.

chapter nine

Altering Constants and Rewriting History

'The first thing to realise about parallel universes . . . is that they are not parallel. It is important to realise that they are not, strictly speaking, universes either, but it is easiest if you try and realise that a little later, after you've realised that everything you've realised up to that moment is not true.'

Douglas Adams[1]

RIGID WORLDS VERSUS FLEXI WORLDS

'Tomorrow I will seven eagles see, a great comet will appear, and voices will speak from whirlwinds foretelling monstrous and fearful things – This universe never did make sense; I suspect it was built on a government contract.'

Robert Heinlein[2]

What is one to make of this strong anthropic idea? Can it be any more than a repackaging of the statement that our complex form of life is very sensitive to small changes in the values of the constants of Nature? And what are these 'changes'? What are these 'other worlds' where the constants are different and life cannot exist?

One plausible view of the Universe, is that there is one and only one way for the constants and laws of Nature to be. Universes are difficult tricks to bring off and the more complicated they are the more pieces there are that need to fit together. The values of the constants of Nature are thus a jigsaw puzzle with only one solution and this solution is completely specified by the one true theory of Nature. If this were true then it would make no more sense to talk about other hypothetical universes in which the constants of Nature take different values than it would to talk of square circles. There simply could not be other worlds.[3] The fact that the one and only possible universe was such as to allow life to develop and persist would be just a brute fact about the world, albeit an extremely agreeable one.[4]

Saddled with this 'rigid world' view we would be unable to say anything further about seemingly fortuitous values of the constants of Nature. In the future we could only wait and watch as a sequence of experimenters checked to more and more decimal places that the values of all the constants of Nature were just as predicted. A rigid world offers no scope for things to be other than what they are; when it comes to the basic laws, forces and constants of Nature,[5] there are no alternatives.

By contrast, the 'flexi world' view offers scope for variation. If there are (or can be) 'other' universes, if some of the constants of Nature are not rigidly specified by the final theory, or if our own Universe displays very different structures beyond our horizon, then the strong anthropic principle has a clear meaning.

Suppose that universes exist in which the constants of Nature take on a wide range of different values. Then there is a collection of different possibilities against which to judge the position of our observed suite of constants. This is what Carter envisaged as a way of transforming an application of the strong anthropic principle into one that amounted to just an application of the weak principle. For if many (or even all) possible universes 'exist' in some sense, then somewhere within the whole constellation of possible combinations of the values of the constants will fall out situations that permit observers to evolve. Inevitably, we live in one of these universes, no matter how special its properties might appear to be when viewed over the entire spectrum of possibilities. Thus, Carter proposes that

> 'It is of course always philosophically possible – as a last resort, when no stronger physical argument is available – to promote a *prediction* based on the strong anthropic principle to the status of an *explanation* by thinking in terms of a "world ensemble". By this I mean an ensemble of universes characterised by all conceivable combinations of initial conditions and fundamental constants . . . The existence of any organism describable as an observer will only be possible for certain restricted combinations of the parameters. A prediction based on the strong anthropic principle may be regarded as a demonstration that the feature under consideration is common to all members of the cognizable subset.'[6]

The idea of there being other universes is not a new one. There was speculation about the possibility in the eighteenth and nineteenth centuries as part of the debate about life on other worlds. There was also considerable discussion in a context very similar to that of the strong anthropic principle. Similar life-supporting coincidences involving the form of the laws of gravity and motion, the constitution of the Earth and the solar system, and human biology had been known for a long time. Natural theologians argued that they showed evidence of Divine purpose in the structure of our Universe. Others, beginning with Leibniz, argued that we lived in the best of all possible worlds – a view mercilessly parodied by Voltaire in *Candide*. However, the perspective changed when Maupertuis showed, with considerable help from the great Swiss mathematician Leonard Euler, that the known laws of motion that Newton had proposed could be derived from a new mathematical principle. The principle allowed one to consider actual motions between two points taking all possible paths. If one evaluated a particular quantity, called the 'action', for each path, and required the actual path taken to have the least value of the action, then this ensured that the path was identical to that predicted by Newton's laws. Eventually physicists found that all laws of physics could be derived from 'action principles' of this form. Maupertuis proudly announced that he could tell what the 'best' of all possible worlds meant and what the other worlds were: 'best' meant least action and the other inferior worlds are those where motion does not follow least action paths. Indeed, during the nineteenth century there was even an attempt to explain fossils as relics of these failed worlds of non-minimal action. By the end of the nineteenth century the evident vastness of the astronomical Universe made it easy to speculate that elsewhere there should be worlds governed by natural laws different from our own. Wallace, writing in 1903, argues that

> 'no two stars, no two clusters, no two nebulae are alike. Why then should there be other universes of the *same* matter and subject to the *same* laws . . . ? Of course, there may be, and probably are, other universes, perhaps of other kinds of matter and subject to other laws.'[7]

Modern physics is built around the derivation of the laws of Nature from action principles. It is the most efficient way to find them and allows far greater generalisation and unification of different laws. Max Born, one of the pioneers of quantum mechanics, foresaw that the quest for a Theory of Everything would become a search for the appropriate least action path through the space of all possibilities:

> 'We may be convinced that [the universal formula] will have
> the form of an extremal principle, not because nature has a
> will or purpose or economy, but because the mechanism of
> our thinking has no other way of condensing a complicated
> structure of laws into a short expression.'[8]

Today, as physicists have followed this path towards deeper and more universal theories of the forces of Nature they have moved steadily towards the flexi world view. There *do* seem to be constants of Nature that are not absolutely fixed by an all-encompassing Theory of Everything. Some appear there but are allowed to take a whole continuous range of values. Others don't appear explicitly in the Theory of Everything at all but emerge at particular stages in the evolution of the Universe by a random process, like a perfectly balanced needle that falls in some particular direction. These constants take on values which manifest the way in which the outcomes of the laws of Nature need not possess the symmetries of the laws themselves: they are far more complicated and haphazard.

One of the big questions facing physicists today is the determination of just how many of the defining constants of Nature will be uniquely and completely specified by a Theory of Everything like the current favoured superstring theory, called 'M Theory'. Those that are omitted from this determination will be allowed to take all sorts of different values without affecting the inner logic and self-consistency of the Theory of Everything. They could have been different if particular sequences of events that had led to their appearance in the early stages of the Universe had developed differently. The nearest we could ever come to an explanation of their values would be by application of an anthropic argument. Maybe all the values available to these constants are equally probable. Nevertheless, we would not be observing unless they fell within the narrow band of values that allows observers to exist.

INFLATIONARY UNIVERSES

'The government admitted for the first time yesterday that
genetically modified farm crops contaminate normal crops
no matter how far apart they are.'

Sarah Schaefer[9]

There are several striking properties of the astronomical Universe which
appear to be crucial in permitting life to develop in the Universe. These
are not constants of Nature in the sense of the fine structure constant
or the mass of an electron. They include quantities that specify how
lumpy the Universe is, how fast it is expanding and how much matter
and radiation it contains. Ultimately, cosmologists would like to explain
the numbers that describe these quantities. They might even be able to
show that these astronomical 'constants' are completely determined by
the values of constants of Nature like the fine structure constant.

Some numbers that define our Universe

- The number of photons per proton
- The ratio of dark to luminous matter densities
- The anisotropy of the expansion
- The inhomogeneity of the Universe
- The cosmological constant
- The deviation of the expansion from 'critical'

Figure 9.1 *Some key constants that describe our universe and distinguish from others we can imagine that obey the same laws.*

The distinctive features of the Universe that are specified by these astronomical 'constants' play a key role in providing the conditions for the evolution of biochemical complexity. We are now going to look at two of them in more detail because the way in which their unusual values are explicable creates an entirely new perspective on the Universe that provides a plethora of 'other worlds' wherein the Anthropic Principle finds a natural and unavoidable application.

As we look more closely at the expansion of the Universe we find that it is delicately poised, expanding very close to the critical dividing line that separates universes which are expanding fast enough to overcome the pull of gravity and keep going forever from those which will ultimately reverse into a state of global contraction and head towards a cataclysmic Big Crunch at some finite time in the future. In fact, so close are we to this critical divide that our observations cannot tell us for sure what the long-range forecast holds. Indeed, it is the close proximity of the expansion to the divide that is the big mystery: *a priori* it seems highly unlikely to exist by chance. Again, this is not totally unexpected. Universes that expand too fast are unable to aggregate material into galaxies and stars, so the building blocks of complex life cannot be made. By contrast, universes that expand too slowly end up collapsing into contraction before the billions of years needed for stars to form have passed.

Only universes that lie very close to the critical divide can live long enough and expand gently enough for the stars and planets to form. It is no accident that we find ourselves living billions of years after the apparent beginning of the expansion of the Universe and witnessing a state of expansion that lies close to the critical divide (see Figure 9.2).

A second distinctive feature of our Universe is its uniformity. The lumpiness level beyond the scale of galaxies is very small, on average only about one part in one hundred thousand. This is important because if it were significantly larger then galaxies would have rapidly degenerated into dense lumps, and black holes would form long before

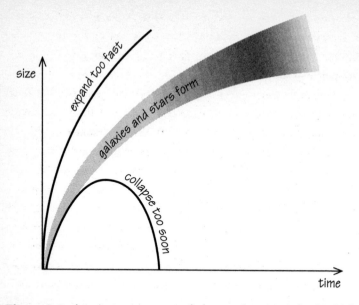

Figure 9.2 *A universe must expand close to the critical divide if life is to evolve. Universes that expand more slowly will collapse before stars have time to form. Universes that expand much faster will not allow material to condense out into islands of matter like galaxies and stars. In neither situation will the building blocks of biochemical complexity be able to form.*

life-supporting environments could be established. Even if they could, the strength of gravity within galaxies would be great enough to disrupt the orbits of planets around stars like the Sun. By contrast, if the lumpiness was much smaller than observed, then the non-uniformities in the density of matter would have been too feeble for galaxies and stars ever to form. Again, the Universe would be bereft of the biochemical building blocks of life: a simpler and less interesting place.

Since 1980 the preferred cosmological theory has provided an explanation for why the Universe displays proximity to flatness, its small (but not too small) level of lumpiness and very large size.[10] These now seem to be features that can be explained by a sequence of events that may be very probable in any type of universe, no matter how it starts out expanding. This theory of the very early Universe introduces an historical interlude called 'inflation'. It creates a slight gloss on the simple picture of an expanding universe. But this gloss has huge implications. The standard Big Bang picture of the expanding universe, that has been with us since the 1920s, has a particular property: the expansion is decelerating. No matter whether the Universe is destined to expand forever, or to collapse back in on itself towards a Big Crunch, the expansion is always being decelerated by the gravitational attraction exerted by all the material in the Universe. The deceleration is simply a consequence of the attractive character of the force of gravity.

It had always been assumed that gravity would ensure that matter and energy would attract other forms of matter and energy. But in the 1970s particle physicists began to find that their theories of how matter behaved at high temperatures contained new forms of matter, called *scalar fields*, whose gravitational effect upon each other could be repulsive.[11] If they were to become the largest contributors to the density of the Universe at some stage in its very early history, then the deceleration of the Universe would be replaced by a surge of acceleration. Remarkably, it appeared that if such scalar fields do exist, then they invariably come to be the most influential constituent of the Universe very soon after it starts expanding, and their influence can be quite brief but decisive. Soon afterwards they should decay without trace into the cosmic sea of ordinary matter and radiation.

The inflationary universe theory proposes that a brief period of accelerated expansion occurs very early in the history of the Universe (see Figure 9.3). This could have occurred because one of the ubiquitous scalar fields came to dominate the density of matter in the Universe. This field then needs to decay quite rapidly. When it does so, its energy heats up the Universe in a complicated way, while the Universe resumes its usual decelerating expansion.

This brief inflationary episode sounds innocuous. But not so: a very short period of accelerated expansion can solve many of our big

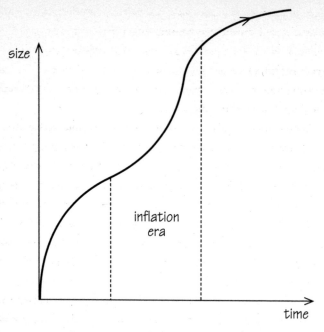

Figure 9.3 *'Inflation' is a brief period of accelerated expansion during the early stages of the universe's history.*

cosmological problems. The first consequence of a short period of accelerated expansion in our past is that it enables us to understand why our visible universe is expanding so close to the critical divide that separates open universes from closed ones. The fact that we are still so close to this divide, after about thirteen billion years worth of expansion, is quite fantastic. Since any deviation from lying precisely on the critical divide grows steadily with the passage of time, the expansion must have begun extraordinarily close to the divide in order to be so close still today (we cannot lie exactly on it[12]). But the tendency of the expansion to veer away from the critical divide is just another consequence of the attractiveness of the gravitational force. It is obvious from just looking at Figure 9.2 that the open and closed universes get farther and farther away from the critical divide as we move forwards in time. If gravity is repulsive and the expansion accelerates, then while this lasts, it will drive the expansion ever closer to the critical divide. If inflation lasted for long enough,[13] it could explain why our visible universe is still so surprisingly close to the critical divide. This life-supporting feature of the Universe would not need to arise from special starting conditions at the Big Bang.

Another by-product of a short bout of cosmic acceleration is that any irregularities in the expansion of the Universe get ironed out and the expansion very quickly goes at the same rate in every direction, just as we see today. This offers an explanation for the extremely symmetric character of the expansion of the Universe, a character trait that has always struck cosmologists as mysterious and unlikely. There are so many more ways to be disorderly than to be orderly that one would have expected a universe pulled out of the hat at random to be a very asymmetrical and disorderly one.[14]

If inflation occurred, the whole visible universe around us today will have expanded from a region that is much *smaller* than it would have originated from had the expansion always decelerated, as it does in the conventional (non-inflationary) Big Bang theory. The smallness of our inflationary beginnings has the nice feature of offering an explanation both for the high degree of uniformity that exists in the overall expansion of the Universe, and for the very small non-uniformities seen by NASA's COBE satellite. These are the seeds that subsequently develop into galaxies and clusters (see Figure 9.4).

If the Universe accelerates, then the whole of our visible universe can arise from the expansion of a region that is small enough for light signals to traverse at very early times. This light traversal enables conditions within that primordial region to be kept smooth. Any irregularities get smoothed out very quickly. In the old, non-inflationary Big Bang theory the situation was very different. Our visible part of the Universe had to emerge from a region vastly bigger than one that light rays can co-ordinate and smooth. It was therefore a complete mystery why our visible universe looks so similar in every direction on the sky to within

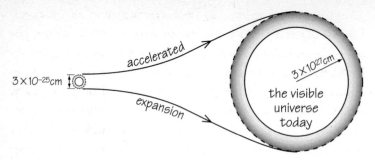

Figure 9.4 *If inflation occurred, the whole visible universe around us today will have expanded from a region that is much* smaller *than it would have originated from if the expansion always decelerated, as in the conventional (non-inflationary) Big Bang theory.*

one part in 100,000, as observations have shown. One part of the visible universe would not have had time to receive light rays from another part far away.

The tiny region which grew into our visible universe could not have started out perfectly smooth. That is impossible. There must always be some tiny level of random fluctuation present. The quantum graininess of matter and energy requires it. Remarkably, a period of inflation stretches these basic fluctuations out so that they extend over very large astronomical scales, where they appear to have been seen by the COBE satellite.[15] In the next year, they will be subjected to minute scrutiny by another satellite (MAP) that was launched in July 2001. If inflation occurred, the signals it receives should have very particular forms. So far, the data taken by COBE is in very good agreement with the predictions, but the really decisive features of the observable signal appear by comparing temperature differences over separations on the sky which are much smaller than COBE can see. The new satellite observations are expected to be made by MAP in 2001 and 2002, and then five years later by the European Space Agency's Planck Surveyor Mission. They will be aided by increasingly accurate observations of smaller portions of sky from the Earth's surface.

In Figure 9.5 we can see a typical prediction of an inflationary Universe model for the form of the fluctuation variation with angular scale, together with the observational data taken by Boomerang near the Earth's surface. Satellite observations will reduce the experimental uncertainties smaller than the thickness of the predicted curve and should provide an inescapably powerful test of particular inflationary cosmological models of the very early Universe. It is remarkable that these observations are providing us with a direct experimental probe of events that occurred when the Universe was only about 10^{-35} seconds old.

Inflation implies that the entire visible universe is the expanded image of a region that was small enough to allow light signals to traverse it at very early times in the history of the Universe. However, our visible part of the Universe is just the expanded image of one causally connected patch approximately 10^{-25} cm in diameter.

Beyond the boundary of that little patch lie many (perhaps infinitely many) other such causally connected patches which will all undergo varying amounts of inflation to produce extended regions of our Universe that lie beyond our visible horizon today. This leads us to expect that our Universe possesses a highly complex geography and the conditions that we can see within our visible horizon, about fifteen billion light years away, are unlikely to be typical of those far beyond it. This complicated picture is called 'chaotic inflation'.[16]

It has always been appreciated that the Universe might have a different structure beyond our visible horizon. However, prior to the investigation of inflationary universe models this was always regarded as an overly positivistic possibility, often suggested by pessimistic philosophers, but which had no positive evidence in its favour. The situation has changed: the chaotic inflationary universe model gives the first positive reason to expect that the Universe beyond our horizon differs in structure to the part that we can see.

It was then realised by two Russian scientific émigrés to the United States, Alex Vilenkin and Andre Linde, that the situation is likely to be even more complicated. If a region inflates then it necessarily creates

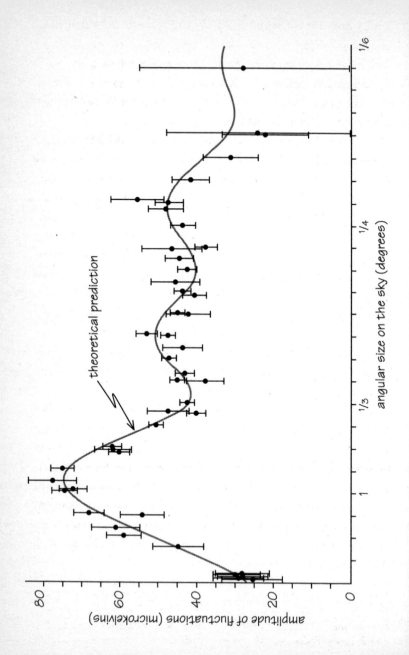

theoretical prediction

amplitude of fluctuations (microkelvins)

angular size on the sky (degrees)

within itself the conditions for further inflation to occur from many sub-regions within. This process can continue into the infinite future with inflated regions producing further sub-regions which inflate, which in turn produce further sub-regions that inflate, and so on . . . *ad infinitum*. The process has no end. It has been called the 'eternal' or 'self-reproducing' inflationary universe[17] (see Figure 9.6).

This enlarged conception of the inflationary model did not set out to produce such an elaborate picture of the Universe. The self-reproducing character of the eternal inflationary universe seems to be an inevitable by-product of the sensitivity of the evolution of a universe to small quantum fluctuations in density from place to place when it is very young.

The eternal and chaotic inflationary structure of the Universe creates a new context for anthropic consideration. In each of the inflated bubbles beyond our visible horizon and all over the past and the future things will have fallen out differently. Each one will have different levels of lumpiness and be closer or farther from the state of critical expansion. It is like picking different universes out of an almost random sample, although they are not really universes, merely extremely large regions bigger than the whole of our observable Universe: 'mini-universes'.

As this scenario has been explored further it has appeared that many more things can be different in each of these inflated bubble mini-universes. They can end up with different numbers of dimensions of space or different constants and forces of Nature. Some of them will not be able to support living complexity of any sort, some will be able to support living complexity of our sort, while others might support life of a completely different sort. Thus, here within our huge, possibly infinite, Universe is the collection of other worlds to which the anthropic principles must be applied.

Figure 9.5 *A typical prediction of an inflationary universe model for the magnitude and variation of the fluctuations with angular separation on the sky, together with the observational data taken by satellites and balloons flown near the Earth's surface.*[18]

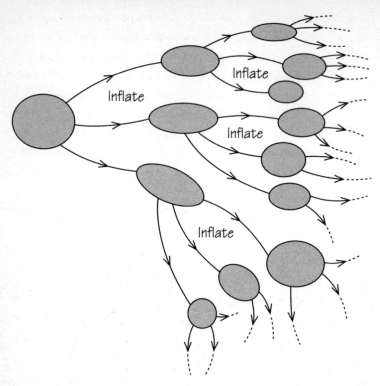

Figure 9.6 *Eternally self-reproducing inflation.*

The challenge that remains for cosmologists is to work out the probabilities for different mini-universes to emerge from this inflationary complexity. Are mini-universes like our own common or rare? Does 'probability' have an unambiguous meaning in this situation? And if life-supporting universes are very rare, what do we conclude? Again, the fact that only a subset of all the possibilities can contain observers is an important consideration when a comparison is made between theoretical predictions and the observed mini-universe. No matter how improbable the life-supporting mini-universes might be, we would have to find ourselves within one.

These considerations bear on interpreting any future quantum cosmological theory. Such a theory, by its quantum nature, will predict that it is 'most probable' that we find the Universe (or its forces and constants) to take particular values. Yet it is not clear that the most probable values will be those we observe. Since only a narrow range of the allowed values for, say, the fine structure constant will permit observers to exist in the Universe, we must find ourselves in the narrow range of possibilities which permit them, no matter how improbable they are. We must ask for the conditional probability of observing constants to take particular ranges, given that other features of the Universe, like its age, satisfy necessary conditions for life. The trend towards unification of apparently independent constants will make the anthropic constraints increasingly severe. In order to test such Theories of Everything we will need to understand all the ways in which the possible existence of observers is constrained by variations in the structure of the Universe, the values of the constants that define its properties, and the number of dimensions it possesses.

VIRTUAL HISTORY – A LITTLE DIGRESSION

'Russia is a country with an unpredictable past.'

Yuri Afanasiev

The little mind game of 'changing' the constants of Nature that the Anthropic Principle provokes us to play has a hitherto unnoticed counterpart in the study of history. There are two aspects to the study of history that will be familiar even from childhood memories of the schoolroom version. There is the need to discover the 'facts' – what happened and when. Following this, there is the need to understand why sequences of events occurred – so as, some suggest, to avoid repeating the mistakes of the past.[19] One response to this two-fold imperative for historical reconstruction has been the creation of 'counterfactual' or 'virtual' history. It might be better named the 'what if?' approach to historical events.

Virtual history[20] tries to predict what might have happened if some pivotal events had not occurred in the past or had been slightly changed. What if the Archduke's motor car had not taken a wrong turning in Sarajevo in 1914? What if Lincoln had not gone to the theatre on the last evening of his life, the votes for Mr Gore and Mr Bush had been fully cast and accurately counted, or Adolf Hitler had been a cot-death victim?

This sounds a little like a parlour game, but attracts surprisingly strong criticism from many historians because it rests upon the assumption of a type of historical determinism that they don't like to admit. As one looks into the issue, it is surprising how many points of similarity there are between the virtual history debate and the discussion of the anthropic impact of varying constants. The contemplation of the consequences of slightly changing the constants of Nature requires the invention of different past histories for the Universe, some of which have the novelty of neither containing ourselves nor any other sentient beings. Cosmologists do not have a full theory that allows all these changes to be incorporated self-consistently but they typically assume that the same laws of change will govern events. Although the basis for the changes in constants or even the 'initial conditions' of the Universe is speculative, the calculation of the consequences can be quite straightforward, a bit like running a computer program with different starting values. By contrast, tinkering with an historical event does not require some change in the laws of Nature, but predicting the outcome is usually too complicated for one to have faith in the results, unless you have the assurance of a novelist.

Historical sequences of events are classic examples of complex systems. They exhibit sensitivity to small changes which make it impossible to predict the future with certainty even though we might be able to understand what has happened in the past. This asymmetry is a feature of all chaotic behaviour, but history is far more unpredictable than a chaotic process. Chaotic processes usually allow one to predict the statistical pattern of future events in a definite way. Historical events have an added sensitivity that renders them unpredictable *in principle* as well as in practice because they involve participants with free will, or at least with the illusion of it.

Weather is hard to predict because it is chaotically sensitive to uncertainties in its present state. But forecasting the weather has no direct effect on the weather. Economic and social forecasters are not so lucky. If a government minister publicly predicts what the economy will do, or a pollster predicts the outcome of an election, these predictions will alter the outcome of what is being forecast in a way that it is logically impossible to include in the original forecast.[21] This is not to say that these events are in some way beyond the rule of logic and intrinsically unpredictable. They can be predicted accurately but that accuracy can only be absolutely guaranteed if the predictions are not made known to the individuals whose actions are being predicted. If they are made known to them, then those individuals can always act to falsify the predictions. These events then become unpredictable in principle, not just in practice.

Virtual histories have become the basis for many fantasies and Hollywood movies, like *It's a Wonderful Life*, in which a suicidal James Stewart is shown how much worse things would have been had he never lived. Alternative outcomes to the Second World War are a favourite scenario for virtual historical novels, notably Kingsley Amis's *The Alteration*,[22] Len Deighton's *SS-GB*,[23] or Robert Harris's *Fatherland*.[24] Often films like *Back to the Future* have used science fiction scenarios of time travel or parallel universes to actualise alternative histories and even bring them into collision with our own. The science fiction scenarios feed on the scientifically possible idea that all possible histories exist. Changing the past just moves the hero, like the wanderer in Jorge Luis Borges's *Garden of Forking Paths*,[25] on to one of the many other historical trajectories that intersect or pass close to the road that would otherwise have been trodden.

The passionate rejection of virtual histories by many historians is very interesting. It is fervent but not very compelling. The philosopher Michael Oakeshott claims that when the historian

> 'considers by a kind of ideal experiment what might have happened as well as what the evidence obliges him to believe did happen [he steps] . . . outside the current of historical thought . . . It is possible that had St. Paul been captured and killed when his friends lowered him from the walls of Damascus, the Christian religion might never have become the centre of our civilisation. And on that account, the spread of Christianity might be attributed to St. Paul's escape . . . But when events are treated in this manner, they cease at once to be historical events. The result is not merely bad or doubtful history, but the complete rejection of history . . . The distinction . . . between essential and incidental events does not belong to historical thought at all; it is a monstrous incursion of science into the world of history . . . The Historian is never called upon to consider what might have happened had circumstances been different.'[26]

The 'monstrous incursion of science' is presumably that of rigid determinism but this is a strange thing to object to. There is no doubt that history was surely a deterministic sequence of events although the sequence may be of such complexity that any hope of connecting all the causes to their consequences is doomed to failure. But commentators like Oakeshott are also concerned that virtual histories tempt us to pick out some facts arbitrarily and make their significance pivotal, whilst viewing others as mere 'accidents'. Benedetto Croce thinks counterfactual history is disastrous[27] for just this reason:

> 'Historical necessity has to be affirmed and continually reaffirmed in order to exclude from history the "conditional" which has no rightful place there . . . What

is forbidden is . . . the anti-historical and illogical "if". Such an "if" arbitrarily divides the course of history into necessary facts and accidental facts . . . and the second is mentally eliminated in order to espy how the first would have developed under its own lines if it had not been disturbed by the second. This is a game which all of us in moments of distraction or idleness indulge in, when we muse on the way our life might have turned out if we had not met a certain person . . . [but] if we went on to such a full exploration of reality, the game would soon be up.'

For these writers, all the historian can do to improve our understanding of what went on in the past is to provide an even more detailed account of events. These critics worry about the division of events into significant and insignificant but have no way of saying which is which except by subjective impression. Nor is there good reason why counterfactual questions should not play a part in interrogating the finished reconstruction of events that is eventually labelled 'history'. This bias is very clear in a revealing account of the historian's aims in the influential little book *What is History?* by the English social historian and historical determinist E.H. Carr:

'From the multiplicity of sequences of cause and effects [the historian] extracts those, and only those, which are historically significant; and the standard of historical significance is his ability to fit them into his pattern of rational explanation and interpretation. Other sequences of cause and effect have to be rejected as accidental, not because the relation between cause and effect is different, but because the sequence itself is irrelevant. The historian can do nothing with it; it is not amenable to rational interpretation, and has no meaning either for the past or the present.'

However, despite this strident opposition to counterfactual rewriting of history, there have been noted historians who have been sympathetic to spinning virtual histories. Gibbon wondered about the subsequent course of European history if the Saracens had not been defeated in the eighth century. In 1907 Trevelyan wrote an essay entitled 'If Napoleon had won the battle of Waterloo' and there have been a host of similar fantasies ever since, feeding on a form of selective causation well displayed by an example of Bertrand Russell's:

> 'Industrialism is due to modern science, modern science is due to Galileo, Galileo is due to the fall of Constantinople, the fall of Constantinople is due to the migration of the Turks, the migration of the Turks is due to the desiccation of Central Asia. Therefore the fundamental study in searching for historical causes is hydrography.'

One contemporary journalist, James Burke, now a *Scientific American* columnist, ran an entire TV series in Britain for a period under the title of *Connections* that traced similarly bizarre causal chains of events.

More serious uses for counterfactual history have been found. Some analysts have attempted to predict the course of economies if certain industries had not developed or if the railways had not existed, in an attempt to discover how much benefit the whole economy gained from specific industries.

To the modern physicist the arguments of idealists like Oakeshott, which deny real accountability to cause and effect, and seem merely to protect their subjects of study from encroachment by others with more rigorous methods, seem wide of the mark. So too do the views of dyed-in-the-wool determinists who see history as an inexorable march towards some inevitable goal of Marxist or capitalist utopia. We understand enough about complex sequences of events to appreciate that it is common for their histories to be predictable in principle but unpredictable in practice because of their sensitivity to small changes, some of which may have gone unnoticed and unrecorded. Thus some past changes to history would have been neutral in their effects, others dramatic. We have also learned that complex systems may display predictable statistical properties, depending upon their detailed character. They may also tend to organise themselves into particular 'critical' states which display a maximum degree of sensitivity to small changes, and it is this state of affairs that allows an overall balance to persist. Remarkably, when it does it is not possible to trace a chain of cause and effect.

There is one area of life where the virtual theory of history is implicit. In the law courts it is often important to judge whether an action resulted in injury. In seeking to establish or cast reasonable doubt upon liability it will be necessary for a barrister to persuade the jury by arguing what would have occurred had the defendant not acted as he did. The prosecutor will create an alternative history in which the accused did not act as he did and try to argue that the sequence of events would have inevitably resulted in no harm occurring. The defending counsel might argue that there is another virtual history in which the harm befell the victim even when his client did not act as he did and so he cannot have been to blame. Such strategies witness to a belief in the importance of virtual histories as a way of testing the stability of particular accounts of history. Of course, identifying the alternative histories is not a guarantee that the truth will appear. Sometimes cause and effect are entwined in a very awkward way. Here is a notorious legal example about the ambiguity of causes:[28]

> 'There is an old story about a man about to cross a desert. He has two enemies. In the night the first enemy slips into his camp, and puts strychnine in his water bottle. Later the same night, the second enemy, not knowing of this, slips into his camp and puts a tiny puncture in the water bottle. The man sets off across the desert; when the time comes to drink there is nothing in the water bottle, and he dies of thirst.

Who murdered him? Defence counsel for the first man has a cast-iron argument: my client attempted to poison the man, admittedly. But he failed, for the victim took no poison. Defence counsel for the second man has a similarly powerful argument: my client attempted to deprive the man of water, admittedly. But he failed, for he only deprived the victim of strychnine, and you cannot murder someone by doing that.'

Historians like Niall Ferguson argue that virtual histories are important. His critics argue that there are an unlimited number of alternatives to consider, which renders reconstruction hopeless. In response Ferguson argues that only a few of the alternative scenarios need be seriously confronted:

'only those alternatives which we can show on the basis of contemporary evidence that contemporaries actually considered.'[29]

Obviously, the reasonable alternatives played a role in the thinking of the protagonist. They would have been his hypothetical futures. How they appeared would have been a factor in the choice of action that was made and therefore essential to our account if we are fully to understand why that choice was made.

This excursion into the philosophy of history aims to show that it is engaged in a lively debate that is curiously analogous to that going on within cosmology about the usefulness of hypothesising universes (or other parts of our Universe) in which the constants of Nature were different from what we find them to be here and now. Virtual natural history is an essential part of modern cosmology.

chapter ten

New Dimensions

'Let us assume that the three dimensions of space are visualized in the customary fashion, and let us substitute a colour for the fourth dimension. Every physical object is liable to changes in colour as well as position. An object might, for example, be capable of going through all shades from red through violet to blue. A physical interaction between any two bodies is possible only if they are close to each other in space as well as in colour. Bodies of different colours would penetrate each other without interference . . . If we lock a number of flies into a red glass globe, they may yet escape: they may change their colour to blue and then be able to penetrate the red globe.'

Hans Reichenbach[1]

LIVING IN A HUNDRED DIMENSIONS

'I am a mathematician to this extent: I can follow triple integrals if they are done slowly on a large blackboard by a personal friend.'

J.W. McReynolds[2]

Arrange to meet someone in a multi-storey complex and you'll need to give them four pieces of information if you want to be sure to meet up at the same time and place. You must specify the time you want to meet, the floor level and the two cross-lanes on that floor: one piece of time information and three pieces of space information. Any less and you may never meet; any more and some of the information will be redundant. These numbers show what it means to live in a universe which has one dimension of time and three of space. Science fiction writers have made a good living out of speculating about extra dimensions which enable us to do magical things in our three-dimensional world by slipping in and out of the visible world. In the nineteenth century there was a famous confidence trickster who claimed to have access to other dimensions so that he could perform 'impossible' feats: unknotting loops of string, turning left-handed spirals into right-handed spirals, moving an object from inside a spherical glass chamber to its outside without penetrating the surface.

To see how stepping into the fourth dimension might help you perform these tricks, think about the jump from two to three dimensions. Place a circular loop of string flat on the table around a lump of sugar. There is no way that the sugar lump can get outside the loop without touching the string if it stays in contact with the flat two-dimensional surface of the table. But if the sugar lump can pass into the third dimension of space this is easily achieved. Just lift it up and put it down again outside the circle of string. Similarly if you put a right-handed spiral coil flat on the table there is no way that it can be changed into a left-handed one by just moving it around on the two-dimensional world of the table-top. But if we lift it into the third dimension and turn it over then it is possible to change the handedness of the spiral (see Figure 10.1).

Despite this fascination with the unseen realms of matter and spirit, there was very little motivation for eighteenth- and nineteenth-century scientists to think about the dimensionality of space. Only one deep thinker seems to have latched on to the deep connection that exists between the number of dimensions of space and the forms of the laws of Nature and the constants that appear within them.

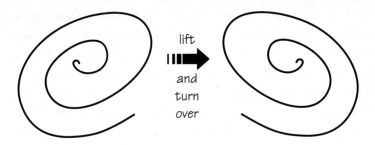

Figure 10.1 *Changing the handedness of a flat spiral by rotating it through the third dimension of space.*

The great German philosopher, Immanuel Kant was far more interested in science than philosophy during his early career in Königsberg (see Figure 10.2). He was a great admirer of Newton and his laws of gravity and motion, and applied himself to understand and apply them to great astronomical problems like the origin of the solar system. As Kant pondered the significance of the special form of Newton's law of gravity he came to ask himself a question that had not been asked before:[3] 'why does space have three dimensions?'

Kant had noticed a very profound thing: that Newton's famous inverse-square law of gravity[4] was intimately connected with the fact that space has three dimensions. If space had four dimensions then gravity would vary as the inverse-cube of distance, if it had 100 dimensions then as an inverse 99th power of distance. In general, an N-dimensional world exhibits a force law for gravity[5] which falls off as the $(N-1)$st power of distance.[6] By the same token, the constant of Nature that appears as the constant of proportionality in these laws will have a value that is determined in part by the numbers of dimensions of space.

Kant used this observation to 'prove' to himself that space must have three dimensions because of the existence of Newton's inverse-square law of gravitational force. He suggested that if God had chosen

Figure 10.2 *Immanuel Kant (1724–1804).*[7]

an inverse-cube rather than an inverse-square law of gravitational force with distance, then a universe of different dimensions – four – would have resulted. Today we would regard this as getting the punch-line back to front: it is the three-dimensionality of space that explains why we see inverse-square force laws in Nature, not vice versa.

Kant's insight showed for the first time that there is a connection between the number of dimensions of space and the forms of the laws of Nature and the constants of Nature that live within them.

Kant went on to speculate about some of the theological and geometrical aspects of extra dimensions, and saw that it might be possible to study the properties of these hypothetical spaces by mathematical means:

'A science of all these possible kinds of space would undoubtedly be the highest enterprise which a finite understanding could undertake in the field of geometry . . . If it is possible that there could be regions with other dimensions, it is very likely that a God had somewhere brought them into being. Such higher spaces would not belong to our world, but form separate worlds.'[8]

His speculation was correct. During the nineteenth century mathematicians 'discovered' other geometries which described lines and shapes on curved surfaces.[9] It was lucky they did. It ensured that Einstein had this 'pure' mathematics available for use when he developed his new theory of gravitation, the general theory of relativity, between 1905 and 1915.

WALKING WITH PLANISAURS

'Mathematics may explore the fourth dimension and the world of what is possible but the Czar can be overthrown only in the third dimension.'

Vladimir Ilyich Lenin[10]

Dimensions are rather important. There are big differences between worlds of different dimensions. One of the simplest is that in two dimensions closed curves divide the world into an inside and an outside. This simple inside-outside result is very important. It makes life rather fractious for a two-dimensional being with a tubular digestive system. If a flatlander tells you his life is falling apart you need to take him seriously as we can see in Figure 10.4.

Stepping up from two to three dimensions also makes mathematicians' lives much more interesting. Paths can meander in very complicated ways in more than two dimensions without intersecting

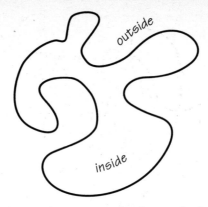

Figure 10.3 *In two dimensions a closed curve divides space into an inside and an outside.*

Figure 10.4 *A two-dimensional being with a digestive system is liable to fall apart.*

(see Figure 10.5). Play higher-dimensional Monopoly, switching to other boards when you land on the stations, or three-dimensional chess, like Mr Spock, and the options grow dramatically.

In fact, three dimensions are the smallest number in which you can get lost. If you walk at random in two dimensions, taking steps of the same size in randomly chosen directions, like a drunkard, then you will eventually return to your starting point. But if you walk at random in three (or more) dimensions of space you will never return to your starting point. You will be lost in space. There are just too many wrong turnings for the random walker to take.

These examples suggest that things always get more complicated as we go from two to three dimensions and higher. But it ain't necessarily so. Sometimes the extra dimensions just make it harder to fit

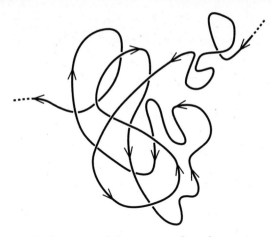

Figure 10.5 *Paths can wind in very complicated ways in more than two dimensions without intersecting.*

things in. Geometers since Plato have recognised that something strange happens as we go from two dimensions to three dimensions. There are an infinite number of regular (equal-sided) polygons in two dimensions but only five regular three-dimensional polyhedra: the famous Platonic solids (see Figure 10.6). The symmetry required to create such solids is very demanding and very few shapes can fit into the three-dimensional space. With more dimensions than three, things become more restrictive still.

The Victorians were strangely beguiled by other dimensions. They saw fantasies about life in fewer or more dimensions as parables through which to make points about our three-dimensional existence. Although these fables are often geometrically interesting, this was rarely their true purpose. What better way for a religious apologist to deflate scepticism about the spiritual realm than to show how blissfully ignorant flatlanders can be of the third dimension that is so plain to us? How better for the illusionist to 'explain' their tricks than by invoking another dimension?

The most famous of fantastic fables, *Flatland: A Romance of Many Dimensions by a Square* was written in 1884 by Edwin Abbott, the

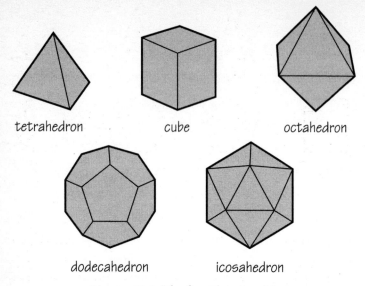

tetrahedron   cube   octahedron

dodecahedron   icosahedron

Figure 10.6 *The five Platonic solids.*

Headmaster of the City of London School. It was a thinly veiled social commentary. The flatlanders[11] and their high priests persecuted anyone who mentioned the unseen third dimension. The more sides people have, so the higher their social rank. Thus, women are lines, the nobility are polygons, and the high priests are circles (see Figure 10.7). The hero is Mr Square who conforms to society's rigid structure until he receives a visit from Lord Sphere from the third dimension, who then propels him into that third dimension to provide a fuller perspective on the nature of reality.[12]

Not everyone was thinking about fewer dimensions. Just a few years before Abbott's book appeared, London society had been rocked by the 1877 trial of the notorious psychic, Henry Slade, who was eventually found guilty of fraud. Some scientists had come to his defence when he claimed to be in contact with the fourth dimension[13] and looked at his claims to emanate objects from it.[14] The occult was a fashionable idea in Victorian England. Even Arthur Conan Doyle

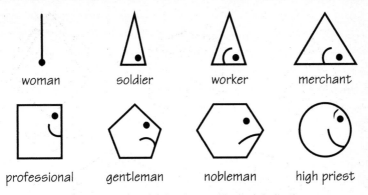

woman soldier worker merchant

professional gentleman nobleman high priest

Figure 10.7 *Some of Edwin Abbott's Flatlanders.*

seems to have believed in fairies.[15] I doubt that Sherlock Holmes did though.[16]

In 1877 a number of simple controlled experiments were set up to test Slade's claims that he could send objects in and out of the fourth dimension:

- Given two unbroken wooden rings, interlock them without breaking the rings.
- Transform a right-handed spiral snail shell into a left-handed one.
- Make a knot in a closed loop of rope without cutting it.
- Given a rope tied in a right-handed knot, inside a sealed container, untie and retie the rope as a left-handed knot without breaking the seal.
- Remove the contents of a sealed bottle without breaking it.

All these tests were devised using mathematical properties of two or three dimensions. The only way to remove the contents of the bottle or unknot the knot is to pass into a higher dimension. As you see, Slade was a sort of nineteenth-century Uri Geller. Alas, he didn't succeed in performing these topology-defying feats under controlled conditions and was eventually found guilty of fraud by the courts.

POLYGONS AND POLYGAMY

'It seems to me that the subject of higher space is becoming felt as serious . . . It seems also that when we commence to feel the seriousness of any subject we partly lose our faculty of dealing with it.'

Charles Hinton[17]

The curious English mathematician, Charles Hinton, worked at the US patent office Washington DC at the same time as Einstein worked in the Swiss patent office. His progressive father, James, had been a surgeon[18] and a charismatic religious philosopher preaching free love and open polygamy; not a recipe for advancement in Victorian England. But young Charles seemed more interested in polygons than polygamy. After studying at Rugby School and Oxford he became a mathematics teacher at Cheltenham Ladies' College and then at Uppingham School. His first published essay 'What is the fourth dimension?' appeared in 1880.[19] Thereafter his life became breathlessly exciting. He had clearly listened to the advice of his father because in 1885 he was arrested for bigamy. He had married Mary Boole, widow of George Boole, one of the creators of logic and set theory, but then married Maude Weldon as well! Imprisoned for three days, on his release he left for the USA with Mary, was hired as instructor at Princeton, and invented the automatic baseball-pitching machine.[20] After being fired from this post, he moved on to the Naval Academy for a period before coming to rest at the US patent office.

Hinton's memorable contribution to the study of higher dimensions was the series of simple pictures that he created to show how it was possible for us to gain a shadowy impression of what four-dimensional objects would look like. He noticed that the pictures that we see in books of real three-dimensional objects are always two-dimensional – flat on the page – and so we should be able to predict what a three- or two-dimensional picture of a four-dimensional object would look like. This image might be its shadow or its projection. Some examples inspired by Hinton are shown in Figure 10.8.

Hinton's ideas for visualising the fourth (and higher) dimensions by extrapolation and analogy were hugely influential and in 1909 the *Scientific American* magazine offered a $500 prize for the best popular explanation of the fourth dimension. In Europe, we see a similar

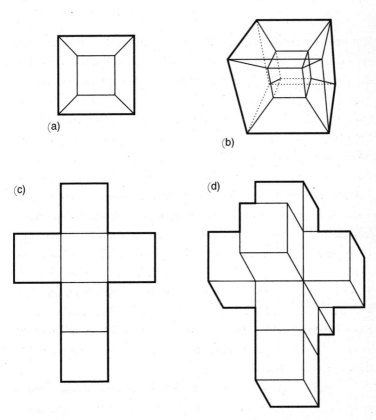

Figure 10.8 *(a) A three-dimensional cube appears two-dimensional when seen in projection. (b) A four-dimensional cube appears three-dimensional when viewed in projection and can be drawn in perspective on the page. (c) Unfolding a cube. (d) Unfolding a four-dimensional cube.*

fascination with multidimensional perspectives emerging in the world of art. The Cubists seized upon the fourth dimension.[21] Marcel Duchamp's *Nude Descending a Staircase* superimposes blurred pictures of a woman as she walks downstairs, visually expressing the fourth dimension of time. A purely spatial ambiguity is exploited by Picasso in his *Portrait of Dora Maar* (see Figure 10.9). Here the idea is to escape the three-dimensional strait-jacket of a single perspective by showing all angles at once when looking at the subject's face.

Figure 10.9 *Pablo Picasso's* Portrait of Dora Maar.[22]

WHY IS LIFE SO EASY FOR PHYSICISTS?

'That was when I saw the Pendulum.

   The sphere, hanging from a long wire set into the ceiling of the choir, swayed back and forth with isochronal majesty.

   I knew – but anyone could have sensed it in the magic of that serene breathing – that the period was governed by the square root of the length of the wire and by π, that number which, however irrational to sublunar minds, through a higher rationality binds the circumference and diameter of all possible circles. The time it took the sphere to swing from end to end was determined by an arcane conspiracy between the most timeless of measures: the singularity of the point of suspension, the duality of the plane's dimensions, the triadic beginning of π, the secret quadratic nature of the root, and the unnumbered perfection of the circle itself.'

Umberto Eco[23]

After you have been using the equations and formulae of mathematical physics for a while you become used to a peculiarity of Nature. It is very forgiving about our ignorance of certain details. The laws of Nature have several ingredients: a logical engine for predicting the future from the present, a place to insert precise information about the present, special constants of Nature, and a collection of simple numbers. These simple numbers show up alongside the constants of Nature in almost every physical formula. In Chapter 3 we saw that Einstein picked them out for Ilse Rosenthal-Schneider and called them 'basic constants'. They are just numbers. For example, the period ('tick') of a pendulum clock is given to high accuracy by a simple formula

$$\text{Period} = 2\pi\sqrt{(L/g)}$$

where L is the length of the pendulum and g is the acceleration due to gravity at the Earth's surface. Notice the appearance of the 'basic constant' $2\pi \approx 6.28$. In every formula we use to describe some aspect of the physical world, a numerical factor of this sort appears. Remarkably, they are almost always fairly close in value to 1 and they can be neglected, or approximated by 1, if one is just interested in getting a fairly good estimate of the result. This is a major bonus because in a problem like the determination of the period of a simple pendulum this allows us to obtain the approximate form of answer. The period, which has dimensions of a time can only depend on the length L and the acceleration g in one way if the resulting combination is to be a time: that combination is the square-root of L/g.

This nice feature of the physical world, that it seems to be well described by mathematical laws in which the purely numerical factors that appear are not very different from 1 in magnitude, is one of the almost unnoticed mysteries of our study of the physical world. Einstein was very impressed by the ubiquity of *small* dimensionless numbers in the equations of physics and wrote of the mystery that, although this almost always seems to be the case,

> 'we cannot require this rigorously, for why should not a numerical factor like $(12\pi)^3$ appear in a mathematical-physical deduction? But without doubt such cases are rarities.'[24]

And many years later in one of his letters to Rosenthal-Schneider about the constants of Nature he remains just as puzzled by this mystery:

> 'It would seem to lie in the nature of things that such basic numbers do not differ from the number 1 in respect of the order of magnitude, at least as long as consideration is

confined to "simple" or, as the case may be, "natural"
formulations.'[25]

It is possible to shed some light on this problem if we recognise
that almost all of the numerical factors that Einstein was so impressed
by have a geometrical origin. For example, the volume of a cube with
edge length R is R^3 but the volume of a sphere of radius R is $4\pi R^3/3$.
The numerical factors allow for the difference in detailed shape when
the forces of Nature are acting. Since the fundamental forces of Nature
are symmetrical and do not have a preference for different directions there
is a tendency towards spherical symmetry. Observations like these enable
us to provide Einstein with a possible answer to his problem.

We know that the perimeter of a circle of radius R has extent
$2\pi R$. The surface area of a sphere is $4\pi R^2$. Likewise, the area of a circle
is πR^2 and the volume of a sphere is $4\pi R^3/3$. Now think about 'spheres'
in N dimensions. Mathematicians can easily calculate what the surface
area and the volume of such spheres will be. It is clear that, $A(N)$, the
area of the N-dimensional ball of radius R will be proportional to R^{N-1}
and its volume, $V(N)$, proportional to R^N, but not at all obvious what
the numerical quantities like '4π' or '$4\pi/3$' will be. The formulae are
shown in the graph of Figure 10.10.

The remarkable feature of the picture is that as the dimension of
space increases so the numerical factors grow to become enormously
different from 1. They don't grow in proportion to N, or even as 2^N.
They grow as N^N. So, we have an answer for Einstein. The ubiquity
of small numerical factors in the laws of Nature and the formulae of
physics is a consequence of the world having a very small number
of space dimensions. If we lived in a world with 20 dimensions then
simple estimates that ignored numerical factors in physical formulae
would be extremely inaccurate in many cases and Einstein would be
asking why they are always so inconveniently large.

From this we see that constants of Nature have a much larger
relative influence when it comes to determining the outcomes of the

a)

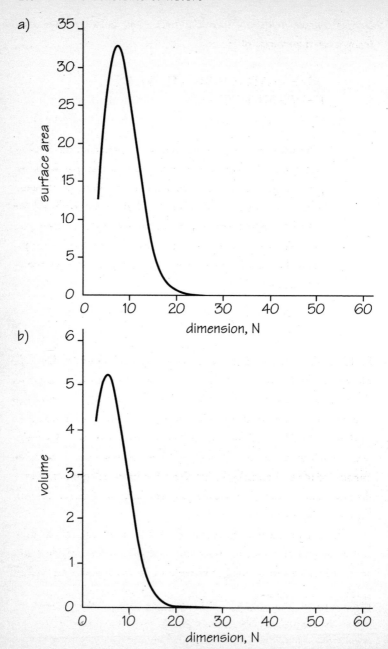

b)

laws of Nature in three dimensions than they do in universes with many more dimensions of space.

THE SAD CASE OF PAUL EHRENFEST

'Ehrenfest was not merely the best teacher in our profession whom I have ever known; he was also passionately preoccupied with the development and destiny of men, especially his students. To understand others, to gain their friendship and trust, to aid anyone embroiled in outer or inner struggles, to encourage youthful talent – all this was his real element, almost more than his immersion in scientific problems.'

Albert Einstein

Paul Ehrenfest was a doubting Thomas; but it was himself he doubted. He was a very talented Austrian physicist who worked with many of the greatest names in science during the early part of the twentieth century: Einstein, Heisenberg, Schrödinger, Pauli, Dirac – all benefited from his help. Above all, he was an incisive critic, able to pick on the weak points of any argument: the conscience of physics. He was also famous for off-beat remarks, like:[26] 'Why do I have such good students? Because I am so stupid.' or 'Do you say that to make a point, or only because it happens to be true?'

Ehrenfest made important contributions to physics in several areas and undergraduates studying quantum mechanics will invariably come across 'Ehrenfest's theorem'. But Ehrenfest's standards were so high that

Figure 10.10 *The variation of the area and volume of a spherical ball in N dimensions with radius equal to one unit of length. The volume has a maximum for N near 5.3 but then falls off rapidly.*

he could not live up to them himself. His childhood had been an unhappy one. His mother died in 1890, when he was 10, and his father, who had suffered chronic bad health, died six years later.

Despite the high esteem in which he was held by others, and which led to his invitation to hold the professorship of physics at Leiden in 1912 when he was only 32 years old (Figure 10.11), Ehrenfest suffered from low self-esteem. He became frustrated by his inability to keep up with the fast pace of developments in quantum physics and their increasingly mathematical nature. In May 1931 he wrote to Niels Bohr that

> 'I have completely lost contact with theoretical physics. I cannot read anything any more and feel myself incompetent to have even the most modest grasp about what makes sense in the flood of articles and books. Perhaps I cannot at all be helped any more.'

His despair deepened, exacerbated by the severe mental problems of his Down's Syndrome son, Wassik. Ehrenfest's famous supervisor, Ludwig Boltzmann, had committed suicide in 1906, despairing at the lack of recognition of his work. Paul Ehrenfest did the same on 25 September 1933, shooting himself, after first shooting his son, in the doctor's waiting room. His last letter of explanation to his closest scientific friends and his students was never sent.[27]

Ehrenfest is part of our story because, in 1917, he was the first to notice[28] how many aspects of physical laws were strongly dependent upon the number of dimensions of space. Building upon Kant's insights into the link between the inverse-square law of gravity and the dimensions of space, Ehrenfest noticed that it was only possible to have planets moving around a central mass (like the Sun) in stable orbits if the world had three dimensions. Following this down to the scale of atoms, where the inverse-square law of electricity and magnetism is responsible for the attractive force between the positively-charged

Figure 10.11 *Paul Ehrenfest (1880–1933) with Albert Einstein.*[29]

atomic nucleus and the negatively-charged electrons moving around it, Ehrenfest showed that in worlds with more than three dimensions *no stable atoms could exist at all*. Either the electrons fell along a spiral path into the nucleus or they dispersed.

Ehrenfest also noticed that three-dimensional waves have very special properties. Only in three dimensions do waves travel in free space without distortion or reverberation. If the number of dimensions of space is *even* (2, 4, 6, . . .) then different parts of a wavy disturbance will travel at different speeds. As a result, if the wavy emission is continuous there will be reverberation at the receiver: waves that left at different times will arrive at the same time. If the number of dimensions of space is an odd number all the disturbances travel at the same speed but if there are not *three* dimensions the wave will become increasingly distorted. Three-dimensional waves are special.

Ehrenfest's imaginative study showed that the dimensionality of the world has a far-reaching effect upon the way things are. Three-dimensional worlds are very unusual.[30] They impose special properties on the laws and constants of Nature.

Yet, in 1917, Ehrenfest went no further and drew no special philosophical conclusions from his results. He was not the first to notice that there was something special about planetary orbits in three-dimensional worlds. William Paley had spelt out the unique life-supporting features of the inverse-square law of gravity back in 1802, and Wallace's 1905 survey of *Man's Place in the Universe* had reiterated these special features. But these authors had written before the quantum theory of matter had emerged and Ehrenfest was able to make a much fuller and deeper case for the physical uniqueness of three-dimensional worlds.

THE SPECIAL CASE OF GERALD WHITROW

'The universe is real but you can't see it. You have to imagine it.'

Alexander Calder[31]

The direct anthropic link between the number of dimensions of space and the existence of living observers was first made by the English cosmologist Gerald Whitrow, in 1955. Asking the question 'Why do we observe the Universe to possess three dimensions?' he sought to provide a new type of answer[32] by arguing that thinking observers could only exist in three-dimensional worlds. Indeed, he suggested that it would be possible to deduce the dimensionality of the world from the fact that we, or another form of intelligent life, exist:

'this fundamental topological property of the world . . . could be inferred as the unique natural concomitant of certain other contingent characteristics associated with the evolution of the higher forms of terrestrial life, in particular of Man, the formulator of the problem.'

He elaborated his arguments in a popular book on cosmology published four years later[33] and attempted to eliminate the possibility of a life-supporting two-dimensional world by arguing that the inevitable intersections of connections between nerve cells in two dimensions would short-circuit the creation of a complex neural network.

Whitrow's approach is the first application of what would now be called the 'Anthropic Principle'. It is earlier than Dicke's application of it to the problem of varying G and the Large Numbers Hypothesis. Using what we know today, we can enlarge it a little further. And if we are going to contemplate what the world might be like if its laws stayed the same but the number of dimensions of space were different, why stop there? Why not ask what would happen if the number of dimensions of time were different as well?[34]

The possibility of universes with different dimensions of both space and time has been explored by a number of scientists.[35] Just as when we considered universes with other dimensions of space and one dimension of time, we can assume that the laws of Nature keep the same mathematical forms but permit the numbers of dimensions of space and time to range freely over all possibilities. The situation is summarised in the picture shown[36] in Figure 10.12.

The chequer-board of all possibilities can be whittled down dramatically by the imposition of a small number of reasonable requirements that seem likely to be necessary for information processing, memory, and therefore 'life', to exist. If we want the future to be determined by the present then we eliminate all those regions of the board marked 'unpredictable'. If we want stable atoms to exist along with stable orbits of bodies (planets) around stars then we have to cut out the strips marked 'unstable'. Cutting out worlds in which there is only faster-than-light signalling we are left with our own world of 3 + 1

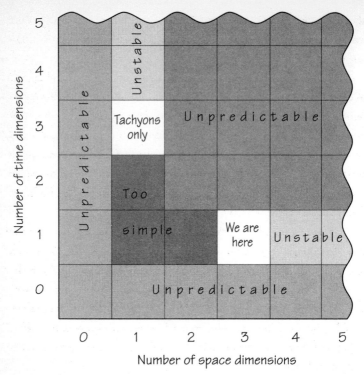

Figure 10.12 *The properties of universes with different numbers of dimension of space and time. One time and three dimensions of space seem to have special properties that are necessary if structures as complex as living beings are to exist.*

dimensions of space plus time along with very simple worlds that have 2 + 1, 1 + 1, and 1 + 2 dimensions of space plus time. Such worlds are usually thought to be too simple to contain living things. For example, in 2 + 1 worlds there are no gravitational forces between masses and there is an imposed simplicity of designs that challenges any attempt to evolve complexity.

Notwithstanding these limitations, there has been much speculation about how working devices could be constructed in two-dimensional worlds.[37] We have already mentioned Whitrow's concerns about producing adequate neural complexity in a two-dimensional world. Networks are extremely limited because paths cannot cross without intersecting.[38]

Worlds with more than one time are hard to imagine and appear to offer many more possibilities. Alas, they seem to offer so many possibilities that the elementary particles of matter are far less stable than in worlds with a single time dimension. Protons can decay easily into neutrons, positrons and neutrinos, and electrons can decay into neutrons, antiprotons and neutrinos. The overall effect of extra time dimensions is to make complex structures highly unstable unless they are frozen in conditions of extremely low temperature.[39]

When we look at worlds with dimensions of space and time other than 3 plus 1 we run into a striking problem. Worlds with more than one time dimension do not allow the future to be predicted from the present. In this sense they are rather like worlds with no time dimension. A complex organised system, like that needed for life, would not be able to use the information gleaned from its environment to inform its future behaviour. It would remain simple: too simple to store information and evolve.

If the number of dimensions of space or time had been chosen at random and all numbers were possible then we would expect the number to be a very large one. It is very improbable that a small number is chosen. However, the constraints imposed by the need to have 'observers' to talk about the problem mean that not all possibilities are available and a three-dimensional space is forced upon us. All the alternatives will be barren of life. If scientists in another universe knew our laws but not the number of dimensions we lived in, they could deduce the number from the fact of our existence alone.

In summary, we have seen that Whitrow's approach to the problem of why space has three dimensions leads to a far-reaching appreciation of how and why three-dimensional worlds with a single arrow of time are peculiar. The alternatives are too simple, too unstable or too unpredictable for complex observers to evolve and persist within them. As a result we should not be surprised to find ourselves living in three spacious dimensions subject to the ravages of a single time. There is no alternative.

THE STRANGE CASE OF THEODOR KALUZA AND OSKAR KLEIN

'The dogmas of the quiet past are inadequate to the stormy present. The occasion is piled high with difficulty, and we must rise with the occasion. As our case is new, so we must think anew, and act anew. We must disenthrall ourselves.'

Abraham Lincoln

Theodor Kaluza (1885–1954) was the only child of a family of scholars that had lived in what was then[40] the German town of Ratibor for more than three centuries. His father Max was a renowned scholar of English language and literature but Theodor showed an early talent for mathematics and was enrolled as a student at the University of Königsberg where he completed a doctorate in 1910. So far, the young Kaluza seemed to be on course for a career as a successful university professor and researcher. He was a friendly man of wide interests, with a keen sense of humour, who spoke and wrote in fifteen languages, but was clearly not much given to the practicalities of life. His son tells us that it was characteristic of his approach to practical matters that in his early thirties he decided that it was necessary for him to learn to swim. He obtained a book about swimming, read it carefully, jumped into the water and swam successfully on his first attempt. Such, he claimed, was the power of theoretical knowledge!

But for some reason Kaluza's career stalled. Instead of remaining, like other talented young scientists, for just two or three years in an assistant position at the university, he stayed there for twenty years, never achieving promotion to a professorship. It was during this lengthy apprenticeship that he decided to write to Einstein about his new ideas linking electricity, magnetism and gravity. The time was April 1919 and Einstein was already renowned amongst physicists for his work on relativity, gravity and atomic physics, although he was not yet a household name amongst the general public. Kaluza had noticed that by adding an extra dimension of space to the world it becomes possible to unite Einstein's theory of gravity with Maxwell's theory of electricity and magnetism in a very economical way. Einstein took a long time to respond to Kaluza's letter but eventually he replied with enthusiasm, urging Kaluza to prepare the work for publication. Einstein added his imprimatur by communicating the work to the *Journal of the Prussian Academy*[41] in December of 1921.

Kaluza's idea was certainly dramatic. Electromagnetism, he claimed, was really just like gravity propagating in an extra dimension of space. But while the theory was mathematically very elegant it had to face up to the awkward question: 'if there is an extra dimension of space why don't we feel its effects?' Kaluza didn't address this awkward question at all.

An answer to this conundrum was provided in 1926 by the Swedish mathematical physicist Oskar Klein (1894–1977), one of Kaluza's former students. Klein had been developing rather similar ideas to Kaluza but had put them aside when he saw that Kaluza had beaten him to it. He had written to Niels Bohr that 'the origin of Planck's constant may be sought in the periodicity of the fifth dimension'.[42] It was simple. The extra dimension of space is extremely small and circular (about 10^{-30} centimetres in circumference) and so its presence is imperceptible. The fine structure constant of Nature that we see in three dimensions takes a numerical value that is controlled by the size of the extra dimension. This Kaluza-Klein theory, as it became known, was of interest for a while but then fell into the background until the 1980s when it re-emerged to become a focus of interest for physicists.

The theory of Kaluza and Klein showed physicists how the world could have extra dimensions of space without falling foul of the problems that Ehrenfest and others had shown were endemic to worlds with more than three dimensions. The trick was simply that dimensionality had to be undemocratic: there could be more than three dimensions of space but they had to be small and unchanging if they were to avoid altering the character of the world that we experience. Forces of Nature must not democratically propagate their influences in all the dimensions: the extra dimensions of space had to be very much smaller in extent than the three we are familiar with.

In the 1980s, physicists started to resurrect the ideas of Kaluza and Klein to see if by adding yet more dimensions it might be possible to join the strong and weak forces of Nature together with electromagnetism and gravity. If this idea could be made to work then the constants of Nature that described the strengths of these forces would be determined by the size of each of the dimensions responsible. For a while it looked as though this novel idea might just work. Serious attempts were made to calculate the value of the fine structure constant in theories with extra dimensions.[43] But gradually the flaws began to show. The simple extra dimensions of Kaluza and Klein could not mimic all the complicated properties of the strong and weak forces of Nature nor accommodate the properties of the idiosyncratic elementary particles they governed. Still, the lessons learned from this approach were important and could be applied to the new superstring theories that repaired the defects of the Kaluza-Klein theories, as we shall see. The most important was that when we open our minds to the possibility that the world possesses more than three dimensions of space then the true constants of Nature must live in the total number of dimensions. The shadows of them that we see in our three-dimensional world can be quite different in value and, most striking of all, need not even be constant.

Kaluza did eventually get a professorship, first at Kiel in 1929 and then at Göttingen in 1935, after Einstein wrote to support his nomination. In his recommendation he particularly drew attention to the novelty of his attempt to unite gravity and electromagnetism with extra dimensions.

VARYING CONSTANTS ON THE BRANE

'There are two ways of spreading light: to be the candle or the mirror that reflects it.'

Edith Wharton[44]

The most interesting consequence of adding extra dimensions of space is that it permits the observed constants of Nature to change. If the world really has four dimensions of space then the true constants of Nature exist in four dimensions. If we move around in only three of those dimensions then we will see or feel only three-dimensional 'shadows' of the true four-dimensional constants. But those shadows need not be constant. If the extra dimension increases in size, just as our three dimensions of the Universe are expanding, then our three-dimensional constants will decrease at the same rate. This immediately tells us that if any extra dimensions are changing they must be changing rather slowly otherwise we would not have called our constants 'constants' at all.

Take a traditional constant of Nature like the fine structure constant. If the size of the extra dimension[45] of space is R then the value of the three-dimensional fine structure 'constant', α, will vary in proportion to $1/R^2$ as R changes. Imagine that we are in an expanding universe of four dimensions but we can only move around in three of the dimensions. The forces of electricity and magnetism can 'see' all the four dimensions and we will find that our three-dimensional part of them will weaken as the fourth dimension gets bigger.

We know that if the three-dimensional fine structure constant is changing it can't be changing anywhere near as fast as the Universe is expanding. This is telling us that any fourth dimension must be very different from the others. Klein's idea was that it is both very small and static. Some extra force traps the extra dimensions and keeps them small. If they don't change in size significantly we need not see any of our constants varying today. A possible scenario imagines that the Universe begins with all its space dimensions behaving in a democratic fashion but then some of the dimensions get trapped and remain static and small ever after, leaving just three to become big, expanding to become the astronomical Universe that we observe today (see Figure 10.13).

In 1982, string theorists first suggested a spectacular answer to an old problem: how do you marry the quantum theory of matter to Einstein's theory of gravity. All previous attempts had failed miserably. They invariably predicted that some measured quantity should be infinite.[46] These 'infinities' plagued all theories with only three dimensions of space and one of time. But in 1984 Michael Green and John Schwarz

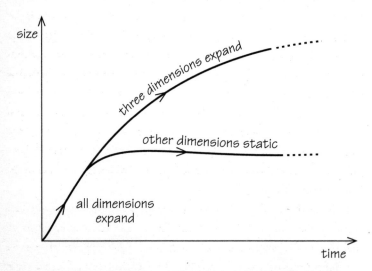

Figure 10.13 *A scenario in which the Universe begins with more expanding dimensions of space than three before undergoing a change into a state where only three continue expanding while the others remain trapped and static.*

showed that this problem could be cured by combining two radical ideas. If one gives up the idea that the most basic entities are point-like, with zero size, and allows there to be more than three dimensions of space, then the infinities miraculously disappear, all cancelling out. As with the earlier Kaluza-Klein theories these extra dimensions can't be changing significantly today or we would see changes in the 'constants' of Nature that govern the structure of our three-dimensional world. Again, they were assumed to be trapped by unknown forces on a very small scale, close to the fundamental Planck length scale of 10^{-33} cm.

The simple idea that only three of the dimensions of space take part in the expansion of the Universe highlights the central mysteries about the dimensions of space and time. We are finding that string theories pick out special numbers of dimensions of space and time together. No reason has been found within those theories to show why only one of the specified number of space-time dimensions is a time; nor why *three* dimensions have become large. If the others are confined to some very small extent then we need to know whether it had to be three dimensions that became large or whether this number fell out at random and could have been different. If the number of spacious dimensions was picked at random by the way events fell out near the beginning of the expansion of the Universe then there might be a different number of large dimensions elsewhere in the Universe beyond our horizon. A random choice would mean that this aspect of the world allowed no further explanation in the normal reductionist sense: only in worlds with three space and one time dimension would we be here to notice the fact.

Recently, another approach to the problem of dimensions and constants has emerged. Rather than simply trap extra dimensions so that they can't change, it allows only gravity to have an influence in all the dimensions of space. The other three basic forces of Nature are confined to act only in three of the dimensions, in a part of the whole Universe that we inhabit called the 'braneworld', see Figure 10.14, so called because it is like a multi-dimensional membrane.

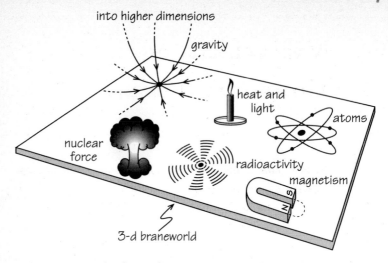

Figure 10.14 *The forces of Nature governing electricity, magnetism, radioactivity and nuclear reactions are confined to a three-dimensional 'braneworld' whilst gravity acts in all the dimensions and is correspondingly weaker.*

The multi-dimensional reach of the force of gravity into the higher dimensions of space, called the 'bulk', is responsible for its relative weakness (by Eddington's famous 10^{40} factor) compared with the other forces that spread their force 'lines' only through three of the dimensions. Braneworlds are subject to intensive paper-and-pencil investigation by physicists all over the world at the moment to see if they can leave some decisive remnant in the Universe which would allow an observational test. In years to come these researches may reveal the link between the constants of Nature that define the true higher-dimensional space in which they exist and the values of their three-dimensional shadows that govern the evolution of the three-dimensional brane that constitutes all that we know of the true Universe. Our constants will be linked to the relative sizes of our braneworld and the unobserved bulk of higher-dimensional space. We may be on the threshold of some deep discoveries that place our entire visible Universe somewhere in hyperspace.

chapter eleven

Variations on a Constant Theme

'A Precambrian physicist would have found it almost easy to build a nuclear reactor.'

George A. Cowan[1]

A PREHISTORIC NUCLEAR REACTOR

'"What do I see landing in the fields nearby but a German plane . . . Two men get out, very polite, and ask me the way to Switzerland . . . one of them comes over to me holding something like a rock in his hand . . . and says, "This is for your trouble; take good care of it, it's uranium." You understand it was the end of the war, . . . they no longer had the time to make the atomic bomb and they didn't need uranium anymore."

"Of course I believe you", I responded heroically. "But was it really uranium?"

"Absolutely: anyone could have seen that. It had an incredible weight, and when you touched it, it was hot. Besides, I still have it at home. I keep it on the terrace in a little shed, a secret, so the kids can't touch it; every so often I show it to my friends, and it's remained hot, it's hot even now."'

Primo Levi, *Uranium*[2]

On 2 June 1972 Dr Bouzigues made a worrying discovery,[3] the sort of discovery that could have untold political, scientific or even criminal implications. Bouzigues worked on the staff at the Pierrelatte nuclear fuel reprocessing plant in France. One of his routine tasks was to measure the composition of ores coming from uranium mines near the river Oklo in the former French colony now known as the West African Republic of Gabon, about 440 kilometres from the Atlantic coast, shown in Figure 11.1. Time after time he checked the fraction of the natural ore that was in the form of the uranium-235 isotope compared to that in the form of the uranium-238 isotope by conducting analyses of uranium hexafluoride gas samples.[4] The difference between the two isotopes is crucial. Naturally occurring uranium that we mine out of the Earth is almost all in the form of the 238 isotope.[5] This form of uranium will not create a chain of self-sustaining nuclear reactions. If it did, our planet would have exploded a long time ago. In order to make a bomb or a productive chain reaction it is necessary to have traces of the active 235 isotope of uranium. In natural uranium no more than a fraction of a per cent is in the 235 form, whereas about 20 per cent is required for a chain of nuclear reactions to be initiated. Weapons-grade or 'enriched' uranium actually contains 90 per cent of the 235 isotope. These numbers allow us to sleep soundly at night secure in the knowledge that the Earth beneath us will not spontaneously begin an unstoppable chain of nuclear reactions that turn the Earth into a gigantic bomb. But who knows, maybe somewhere there is more 235 than average?

Bouzigues measured the 235 to 238 isotope ratios with great accuracy. They were important checks on the quality of materials that would ultimately be used in the French nuclear industry. This was routine work but on that June day in 1972 his attention to detail was rewarded. He noticed that some samples displayed a 235 to 238 ratio of 0.717 per cent instead of the usual value of 0.720 per cent usually found in all terrestrial samples – and even in meteorites and Moon rocks. So accurately was the 'usual' value known from experience[6] and

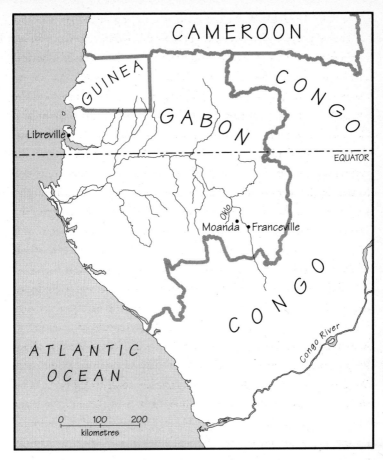

Figure 11.1 *The location of Oklo in West Africa.*

so precisely was it reflected in all the samples taken that this small discrepancy sounded alarm bells. Where was the missing 0.003 per cent of uranium-235? It was as if the uranium had already been used to fuel a nuclear reactor and so the 235 abundance had been depleted before it was mined out.

All sorts of possibilities were considered by the French Atomic Energy Commission. Perhaps the samples had been contaminated by some used fuel from the processing plant? But there was no evidence of any of the intense radioactivity that would accompany spent fuel, and no depleted uranium hexafluoride was missing from the plant's inventory. Some form of terrorist theft of material or extraterrestrial deposit was even suggested. But gradually the investigations found the source of the discrepancy to lie in the natural uranium deposits themselves. There was a naturally low 235 to 238 ratio in the mine site seams. The investigators looked at each step of the uranium ore's transportation and processing, from the original ore mining and local milling in Gabon, to the processing in France before it reached the enrichment plant at Pierrelatte. Nothing untoward was discovered. The uranium from the Oklo mine was just different from that found anywhere else. Indeed, samples that had been kept from all the shipments dispatched to France since the mine began excavating in 1970 all showed slight uranium-235 depletion. Out of the 700 tons of uranium already mined, the total 'missing' mass of uranium-235 amounted to 200 kilograms.

As the mine site was investigated in greater and greater detail it was soon clear that the missing uranium-235 had been destroyed within the mine seams. One possibility is that some chemical reactions had removed it whilst leaving the 238 unscathed. Unfortunately, the relative abundances of uranium-235 and 238 are not affected differently by chemical processes that have occurred inside the Earth. Such processes can make some parts of the Earth rich in uranium ore at the expense of others by dissolving it and moving it around, but they don't alter the balance of the two isotopes that make up the dissolved or suspended ore. Only nuclear reactions and decays can do that (see Figure 11.2).

Gradually, the unexpected truth dawned on the investigators. The depleted seams of uranium-235 contained the distinctive pattern of 30 or more other atomic elements that are formed as by-products of nuclear fission reactions. Their abundances were completely unlike those occurring naturally in the rocks where fission reactions had not occurred. The tell-tale signature of nuclear fission products is known from man-made

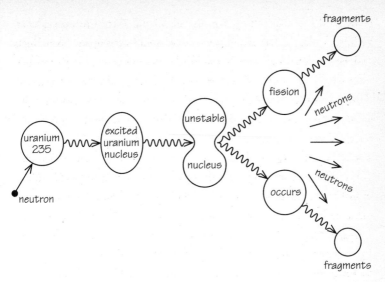

Figure 11.2 *The fission of a uranium-235 nucleus.*

reactor experiments. Six of these distinctive seams of natural nuclear reactor activity were eventually identified at Oklo. Some of the elements present, like neodymium, have many isotopes but not all are fission products. The non-fission products therefore provide a gauge of the abundance of all the isotopes before the natural reactions began and so enable us to determine the effects and running-times of those reactions.[7]

Remarkably, it appeared that Nature had conspired to produce a natural nuclear reactor which had produced spontaneous nuclear reactions below the Earth's surface two billion years ago.[8] It was this episode in the geological history of Gabon that had led to the accumulation of fission products at the site of the present-day mine. As a result of these sensational discoveries, mining was stopped for a period in 1972 while a detailed geochemical survey was carried out. Eventually, 15 fossilised ancient reactor sites were found, 14 of them at Oklo and another about 35 km to the south, at Bangombe.

In 1956, a Japanese physicist named Paul Kuroda, working at Arkansas University, had predicted that just such a thing might happen in Nature.[9] Kuroda considered almost all of the key requirements: the concentrations of uranium needed for nuclear reactions, the time in the past when it might happen and the uranium-235 to 238 ratio.[10] But nowhere could he think of a site on Earth where all these special conditions could be met at once. But Kuroda missed one interesting possibility that the Oklo geology had created for itself.

The first man-made nuclear reactions were produced on 2 December 1942 as part of the famous Manhattan Project that culminated in the creation of the first atomic bombs. They broke heavy nuclei into lighter ones, releasing energy and fast-moving neutrons which went on to break up more heavy nuclei and release yet more energy and neutrons. Man-made reactors are controlled by introducing a 'moderator', like graphite or water, which absorbs neutrons and slows the reaction. Neutrons are emitted with high speeds and in that state are readily absorbed by uranium-238 nuclei. They need to be slowed down considerably in order to have a high probability of being absorbed by another uranium-235 nucleus and so sustain the chain of fission reactions. Rods of graphite can be introduced into the interaction region and retracted when needed to moderate the reactions. Without this moderating effect nuclear reactions would snowball out of control once they have reached a critical level. So what moderated events at Oklo?

Investigators found the distinctive 'smoking gun' of fission products at Oklo, showing that nuclear chain reactions had taken place. Although today the natural abundance of uranium-235 is only about 0.7 per cent relative to uranium-238, the ratio of the two isotopes has not been constant throughout the history of the Earth. They both decay slowly but at different rates. The half-life of 235 is about 700 million years while that of 238 is about 4.5 billion years. The faster decay of 235 means that there was more 235 relative to 238 in the past than there is today. When the Earth formed about 4.5 billion years ago, natural uranium contained about 17 per cent uranium-235. After about 2.5 billion years, when the Earth was 2 billion years old, the 235:238 ratio would have fallen to around 3 per cent, just about right to start a chain reaction that could be moderated by water.

The Oklo uranium deposits were first discovered in the 1960s and are several kilometres long and about 700 metres wide. They derive from uranium originally deposited in the Earth's crust during the formation of the Earth. The original abundance was quite small, on average just a few parts in a million of the Earth's make-up. Its source, like all the other heavy elements in the Earth, lies in the stars. Uranium was formed in the stars and ejected into space before condensing into small rocks that were aggregated into solid planets during the early history of the solar system. Following the intense geological activity associated with the era after the formation of the Earth, the Oklo natural reactors were made possible by the accidental deposition of a uranium-rich seam inside a layer of sandstone lying on top of sheets of granite. Over millions of years nearly a kilometre of sandy sediment was washed down on top of the uranium. The granite layers are tilted at about 45 degrees and this led to a build up of rainwater and soluble uranium oxide deep underground at the bottom of the slope (see Figure 11.3).

The oxidising environment needed to create the water required to concentrate the uranium was brought about by a significant change in the Earth's biosphere. About two billion years ago a change of atmosphere occurred, brought about by the growth of blue-green algae, the first organisms able to carry out photosynthesis. Their activity increased the oxygen content of the water and allowed some of the uranium to change into soluble oxides. At Oklo, the uranium deposits were buried deep enough to prevent them being redissolved and dispersed during nearly two billion years of subsequent history. Only during the last two million years have parts of the ore deposit come close to the surface where it was found by mineral prospectors and mined out.

This is not the end of the special circumstances needed for a natural reactor. The layer of concentrated uranium ore needs to be thick enough to prevent the neutrons created by the first nuclear reactions from simply escaping and it must also be free from contamination by

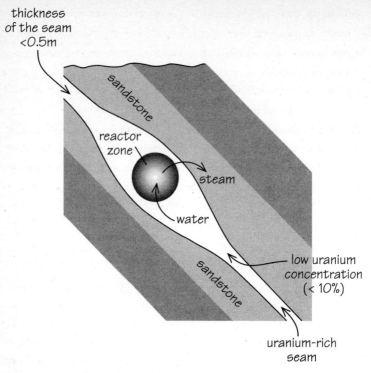

thickness
of the seam
<0.5m

sandstone

reactor
zone

steam

water

sandstone

low uranium
concentration
(< 10%)

uranium-rich
seam

Figure 11.3 *The geology of the Oklo reactor site features granite layers tilted at about 45 degrees. This created a build-up of rainwater and soluble uranium oxide deep underground.*

neutron poisons that will absorb all the neutrons and shut down the chain reactions.

Once the soluble uranium reached a concentration of more than about 10 per cent two billion years ago, nuclear reactions could not only commence but continue in a stable self-regulated fashion. The seams needed to be at least half a metre thick in order that the neutrons did not just escape and the reactions die out. As the reactions went faster, so the temperature increased, turning the water into steam and slowing the neutrons that collided with the water molecules. This slow-down reduced the temperature, causing the steam to condense back into liquid water and reducing the number of neutrons absorbed. As a result the reactions then speeded up. This cycle of stop-go activity seems to have been repeated intermittently over nearly a million years, with episodes of chain reaction lasting for periods varying from just a few years to thousands of years before the reactor finally switched itself off.[11] At six sites within the Oklo uranium layer about a ton of uranium-235 had fissioned away,[12] producing a million times more energy than would have been produced by the long-winded process of natural radioactive decay into uranium-238. At each site, the characteristic pattern of fission decay products remains to tell the tale.[13] This is remarkable enough, but the insights that followed have made the Oklo reactors an important touchstone for our understanding of the constants of Nature.

ALEXANDER SHLYAKHTER'S INSIGHT

'To my mind radio-activity is a real disease of matter. Moreover it is a contagious disease. It spreads. You bring those debased and crumbling atoms near others and those too presently catch the trick of swinging themselves out of coherent existence. It is in matter exactly what the decay of our old culture is in society, a loss of traditions and distinctions and assured reactions.'

H.G. Wells, *Tono-Bungay*[14]

Alexander Shlyakhter was a remarkable young nuclear physicist from St Petersburg (Figure 11.4). He died of cancer in June 2000 after moving to Harvard University in the United States. His expertise was important in the control and understanding of several nuclear accidents, notably the disaster at the Chernobyl reactor in the former Soviet Union. Whilst still a student he realised that the remnants of the past Oklo nuclear activity might be telling us something very important about how nuclear reactions operated two billion years ago. He recognised that there was something very unusual about some of the nuclear reactions involved at Oklo. Remarkably, one of the reactions that occurred there, the capture of a neutron by a samarium-149 nucleus to produce the samarium-150 isotope and a photon of light, is very sensitive. It only occurs because of a fortuitous 'resonance': the dramatic increase in the rate of a nuclear reaction in a particular narrow energy range. This occurrence of a resonance is rather like a hole-in-one in golf. It happens when the energies of the incoming components of a reaction have energies which add up to give a total that is almost exactly equal to the energy state of a possible outcome. In that case the interaction goes through very swiftly into its nicely located final state. It was just the same type of coincidence that Fred Hoyle had predicted should occur in the carbon nucleus which we described in Chapter 8.

Shlyakhter realised that the need for a very precisely located resonant energy level for the capture of a neutron by samarium-149 meant that the Oklo reactor was telling us something very remarkable about the constancy of physics over billions of years. The very finely tuned coincidences that appear to exist between the values of the different constants of Nature which determine the precise energy of this resonance level must have been in place to high accuracy about two billion years ago when the natural reactor was running. In Figure 11.5 we show the probability for the samarium reaction to occur at different temperatures if we shift the present position of the resonance energy. A zero shift means it has the same value as observed in nuclear reactions today.

The resonant character of neutron capture by samarium-149 is responsible for its very significant depletion at the Oklo site. Three of the four forces of Nature, the strong nuclear interaction, the weak interaction and the electromagnetic interaction, play a role in setting the location of the crucial resonance energy level. Unfortunately, the

Figure 11.4 *Alexander Shlyakhter (1951–2000).*[15]

way in which they do so cannot be calculated in full detail because of the sheer complexity of the competing contributions. But Shlyakhter cut through these complexities by making the reasonable estimate that the contribution of each force of Nature to the resonance energy level would be in proportion to its strength. By assuming the temperature of the reactor was about 300 degrees centigrade – the boiling point of the water in the high-pressure environment of the seam – he concluded that 2 billion years ago the resonance level could not have been more than 20 milli electron volts (meV) away from its present position: that is *a change of less than one part in 5 billion over 2 billion years.*

These deductions mean that if the interaction strength between a single neutron and the samarium nucleus is changing then its rate of change is less than 10^{-19} per year, or less than about one part in a

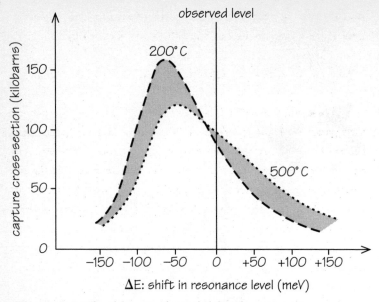

Figure 11.5 *The change in the probability for a samarium nuclear capture reaction to occur at different temperatures as we shift the position of the resonance energy.[16] A zero shift means it has the same value as observed in nuclear reactions today.*

billion over the 14-billion year history of the Universe. Shlyakhter[17] argued[17] that if the interaction strength is determined predominantly by the strong nuclear force then its associated constant of Nature, α_s is subject to the stringent restriction:

{the rate of change of α_s}/{the value of α_s} < 10^{-19} per year

If only the electromagnetic interaction is changing with time then, because its contribution to the total samarium interaction rate is about 5 per cent, any rate of change of the fine structure constant, α, must obey the limit

{the rate of change of α}/{the value of α} < 5×10^{-17} per year

And if only the weak force of radioactivity were to have varied over time, then the variation of its strength, α_w, is bounded by

{the rate of change of α_w}/{the value of α_w} < 10^{-12} per year

These limits were far stronger than any limits on the possible time-variation of the constants of Nature that had ever been found before. The Universe has been expanding for about 14 billion years and so these limits, if taken at face value, are telling us that the fine structure constant cannot have changed by more than about one part in ten million over the entire age of the Universe. Previous observational limits were more than a thousand times weaker.

There are a few things that are immediately clear about these strong limits on the possible variation of the constants of Nature:

(a) They have a particular reach in time back to about 2 billion years ago, when the Oklo reactor formed, compared with 4.6 billion years for the age of the Earth and about 14 billion years for the expansion age of the Universe.

(b) If different constants varied simultaneously then the results might change.

(c) A particular simplifying assumption was made about the way in which the constants of Nature contribute to the neutron capture resonance energy.

(d) Some simplifying assumption has been made about the temperature inside the reactor when it was operating.

The unique probe of the constancy of constants that Oklo provides has ensured that Shlyakhter's brilliant observation has been investigated in much greater detail by others.[18] The most detailed study has been carried out by Yasanori Fujii and his collaborators[19] in Japan. Looking at Figure 11.6, we can see how a shift in the resonance energy (non-zero ΔE_r) plotted along the horizontal axis produces a change in the neutron capture probability, plotted up the vertical axis, that depends upon the temperature of the reactor. The allowed range for the neutron capture probability two billion years ago is between 85 and 97 kilobarns if the abundance of samarium is to agree with the range observed in the reactor sites. The various investigators of the samples agree that the temperature must have been somewhere between 200 and 400 degrees centigrade. Now, one can see from the curves drawn for these temperatures that there are actually *two* ranges of the shift ΔE_r that keep the capture cross-section within the allowed bounds:

$$-12 \text{ meV} < \Delta E_r < 20 \text{ meV}$$

taking the right-hand branch; and

$$-105 \text{ meV} > \Delta E_r > -89 \text{ meV}$$

if we take the left-hand branch.

The limit from the right-hand branch is a refinement of Shlyakhter's original result and leads to a more stringent limit on possible time variation of the fine structure constant if it is the only constant that is assumed to vary. The limit is

$$\frac{\{\text{the rate of change of } \alpha \}}{\{\text{the value of } \alpha \}} = (-0.2 \pm 0.8) \times 10^{-17} \text{ per year}$$

and is about five times stronger than the earlier one. It allows there to be no variation at all because of the ± 0.8 uncertainty in the inferred value. This uncertainty would need to be reduced well below ± 0.2 in order for there to be believable evidence for any actual variation. However, if we take the left-hand branch result then it does not allow ΔE_r to be zero and leads to the deduction that there has been a non-zero change in the value of the fine structure constant since the Oklo event, equal[20] to

$$\frac{\{\text{the rate of change of } \alpha\}}{\{\text{the value of } \alpha\}} = (4.9 \pm 0.4) \times 10^{-17} \text{ per year.}$$

If one looks at the abundances of the other isotopic residues of the Oklo event then this second result might be excluded.[21] But so far the data sample quality and uncertainties about the temperature in the reactor prevent us from ruling it out definitively.

It is also interesting to see the consequences of allowing the electromagnetic and strong nuclear force strengths to vary in time simultaneously. Typically, this leads to limits on the time variation of both 'constants' which are about as strong as those we have just given for the fine structure constant. But there is a peculiar situation, albeit looking rather contrived, in which the limits on variation are far weaker. If, for some unknown reason, the rates of change in the strong and electromagnetic interactions over 2 billion years are equal to within one part in ten million then the effects of the two constant changes cancel. The new limits are dramatically weakened to a level that would have been the case if there was no special neutron capture resonance at all:

$$\frac{\{\text{time rate of change of } \alpha \, [\text{or } \alpha_s]\}}{\{\text{value of } \alpha \, [\text{or } \alpha_s]\}} < 10^{-10} \text{ per year}$$

Although this finely-tuned, one-in-ten-million chance for the possible variation of the electromagnetic and strong force constants might sound rather contrived, it is actually a prediction that they vary at exactly the same rate in a wide range of theories which attempt to join together the different forces of Nature, so this possibility should not be excluded as absurdly unlikely.[22]

THE CLOCK OF AGES

'the first nine digits after the decimal can be remembered
by e = 2.7(Andrew Jackson)^2, or e = 2.718281828 . . . ,
because Andrew Jackson was elected President of the
United States in 1828. For those good at mathematics on
the other hand, this is a good way to remember their
American History.'

Edward Teller[23]

To most people the word radioactivity brings to mind a sentence in
which there also appear words like accident, waste, leak, cancer or
disaster. But without radioactivity we would not be here. The delicate
sequence of processes that create the steady flux of solar energy that
bathes the Earth is made possible by radioactivity. When the Earth
condensed into its present mass of material about four and a half
billion years ago it contained enough metals like nickel and iron at its
core to sustain a significant magnetic field. Without it, we would have
no life-sustaining atmosphere. The wind of electrically-charged particles that are continually blown away from the Sun's surface would have
stripped our atmosphere away, just as they have on Mars where there
is no magnetic shield. The Earth's magnetic field defends us against
these invaders by deflecting them around the atmosphere.

Along with this life-sustaining inner core of iron and nickel, the
primordial Earth also picked up enough radioactive elements, like
uranium, to maintain a long period of heating by radioactive decays
deep inside its interior. This inner engine played a key role in unlock-
ing the Earth's geological potential. The subterranean furnace has stimulated continual editions of mountain building and plate tectonics,
keeping the surface alive and changing in a way that provides a suit-
able habitat for land animals and amphibians.

When the idea that some of the traditional constants of Nature
might be slowly changing was first suggested by Dirac and Gamow,

many physicists realised that the constants that controlled radioactive decay must be crucial for the history of planet Earth. Any change in their past values would most likely upset a delicate balance and create too much or too little heating.

Radioactive elements act as clocks. Their 'half-lives' tell us the time required to halve their initial abundance. They fall into groups with half-lives that are billions, millions and thousands of years respectively.

Following the first attempts by Denys Wilkinson[24] to get limits on the constancy of constants by these means in 1958, Freeman Dyson[25] used the half-life of long-lived beta-decaying nuclei, such as rhenium-187, osmium-187 and potassium-40 to place a limit on possible past variation of the fine structure constant from its present value. These three nuclei have very long half-lives that have been determined accurately by laboratory experiments and by comparison with the ages of meteorites. Given that the uranium-238 decay rate must have been within 20 per cent of its present value over the last 2 billion years, one deduces that

$$\{\text{the rate of change of } \alpha\}/\{\text{the value of } \alpha\} < 2 \times 10^{-13} \text{ per year}$$

Similar studies of different decay sequences by other scientists[26] led to other limits of a very similar strength. These limits were eventually superseded by evidence of the Oklo natural reactor.

UNDERGROUND SPECULATIONS

'This rock salt is over 200 million years old, formed through ancient geological processes in the German mountain ranges. Best before 04 2003.'

Product label[27]

The Oklo phenomenon may well not have been unique. The conditions needed to sustain chains of nuclear fission reactions are unusual but not in any way bizarre. It is possible that other natural reactor sites have been mined out unnoticed or lie awaiting discovery at other sites on Earth. Although there are other sites in Africa and in Colorado, USA, that display deficits of uranium-235 that might have been created by naturally occurring nuclear reactions, none is believed to be a natural reactor.

The discovery of these possible natural reactor sites is important not only for studies of the constants of Nature. They tell nuclear physicists important things about the future stability and confinability of nuclear fission products buried underground for very long periods of time. Maybe one day a piece of very careful chemical book-keeping will lead to a replay of the exciting sequence of investigations that unmasked the Oklo reactor.

If natural reactors can occur on Earth then why not elsewhere? It is tempting to speculate that a new source of life-sustaining heat energy has been identified which might play an unusual role in incubating biochemical evolution on other worlds. The astronomer Fred Hoyle[28] once wrote a science fiction novel about the development of life on a comet that was initiated and sustained by natural nuclear reactions occurring within its core. Perhaps the search for extra-solar planets will discover a planet or a moon on which the Oklo phenomenon occurred on a vaster scale, heating up the interior for long periods of the planet's life and sustaining the development of complex bacterial life, before shutting down and leaving the planet dormant and superficially dead.

It is sobering to think that the time in the history of the Universe when life exists has dictated some interesting nuclear consequences for human life. We have seen how the different decay rates of the two uranium isotopes make uranium-235 relatively more abundant in the past. By the same token it will be relatively less abundant on planets like the Earth in the far future. During the last century we discovered that our planet's crust contains radioactive elements that enable nuclear bombs to be created with some technical skill if we refine the active uranium-235 isotope from the more abundant uranium-238. If humans appeared far earlier or far later on our planet than they did then their prospects for harnessing nuclear weapons would have been very different. Here is the prescient analysis of John von Neumann, one of the most remarkable scientists of the twentieth century, written at the dawn of the nuclear age:

> 'If man and his technology had appeared on the scene several billion years earlier, the separation of uranium 235 [crucial for making bombs] would have been easier. If man had appeared later – say 10 billion years later – the concentration of uranium 235 would have been so low as to make it practically unusable.'[29]

We are the beneficiaries of many aspects of the Earth's interesting geology. The presence of heavy elements with interesting magnetic and radioactive properties has led to our understanding of these fundamental forces of Nature. Life on a pleasant, irrigated planet, bathed in the light of a well-behaved star, would be possible with nothing of nuclear or radioactive interest anywhere near its surface. But its inhabitants would be severely handicapped in their quest to understand the scope and richness of the forces and constants of Nature.

chapter twelve

Reach for the Sky

'An idea that is not dangerous is unworthy of being called
an idea at all.'

Oscar Wilde[1]

PLENTY OF TIME

'All that I know
Of a certain star,
Is, it can throw,
(Like the angled spar)
Now a dart of red,
Now a dart of blue.'

Robert Browning, *My Star*[2]

Imagine that the Son of the Hubble Space Telescope has detected signs
of intelligent life in a star system elsewhere in our Galaxy. Directed
radio signals are beamed out and a reply comes back a few years later.
A slow conversation ensues with each side fairly easily decoding the
incoming messages. Gradually, we learn something odd and slightly disap-
pointing (at least for some people) about our extraterrestrial penfriends
– they are only interested in astronomy. Their civilisation seems to study
nothing else. All developments in mathematics, engineering, computers
and other sciences are subjugated to advancing understanding of the

stars. We don't know quite why this is. Perhaps there is a deep religious imperative. They certainly do other technical things but seem to have little interest in them unless they have cosmic applications.

Whilst terrestrial astronomers are not unhappy to discover this bias, many others are disappointed to have discovered specialists. They decide that one of the things they could best ask their interstellar correspondents about is the values of the constants of Nature. It is not too difficult to make sure we are all talking about the same thing. After all, the radio signals themselves provide an example of shared electromagnetic experience. It isn't too difficult to tell them what we mean by the fine structure constant. The extraterrestrials are asked to measure the ratios of various frequencies of oscillation in atoms and molecules containing specified numbers of particles in and around their nuclei and to send the answers to us at the speed of light. We will do the same and send our answers to them.

As this hasn't happened yet I can't tell you what the comparison revealed. But this little fiction illustrates how information gleaned in other parts of the Universe could give us a unique check on the uniformity of the constants of Nature and the laws of physics. What if we could cut out the extraterrestrials altogether and just gather information about the constants of Nature directly far away in the Universe?

Remarkably, these fictions have been turned into fact without the expense or complications of extraterrestrial communication and decipherment becoming involved. When we observe a distant star we are not only gathering information from far away but also reaching back in time. Light travels at a finite speed and so the farther away a star is from us the longer it has taken for its light to reach us. In the case of the Sun the light travel time is very short, about 8 minutes. The nearest star to us beyond the Sun is Alpha Centauri, 4.1 light years away, while the most distant astronomical objects that are routinely observed are more than 13 billion light years away. The light from these distant objects must be bringing us important information about the physical processes that produced it far away and long ago.

George Gamow was one of the first to have the idea of using astronomical observations[3] in some way to investigate whether constants varied; in fact, he wanted to assume that the fine structure constant *did* vary in a way that would explain Dirac's Large Number coincidences and then see if this change would contribute to the redshifting of light from distant galaxies. The expansion of the universe means that distant galaxies are receding from us and so the light waves their stars emit are received by our telescopes with a lower frequency than they are emitted with. This means that their colours are shifted towards the red end of the spectrum, hence they are 'redshifted'. Gamow saw how to use the redshift to look back in time to see what the constants of Nature were like when the light began its intergalactic journey to our telescopes. In Figure 12.1, we can see Gamow's telegram to his former student, Ralph Alpher, telling of his new idea and some of its implications.

Alas, Gamow's idea turns out to produce no measurable effect even if the fine structure constant varies. But it was not long before three astronomers, John Bahcall, Maarten Schmidt and Wallace Sargent at Cal Tech in Pasadena, hit on another approach that the recent discovery of quasars, or quasi-stellar radio sources, at high redshifts had made possible for the first time.

They had recently found pairs of spectral lines,[4] called 'doublets', created by absorption of light received from the newly discovered quasar QSO 3C191 by the element silicon. The distance between the two lines of the silicon doublet is a small and sensitive feature of atom physics that is a consequence of the relativistic effects that arise when electrons move close to the speed of light around the atomic nucleus (see Figure 12.2). Crucially, the separation of the lines forming the silicon doublet depends sensitively on the value of the fine structure constant.

The quasar 3C191 was located at a redshift of 1.95 and so its light left when the Universe was just one-fifth of its present age, nearly 11 billion years ago, carrying encoded information about the value of the fine structure constant at that time. To the accuracy of the

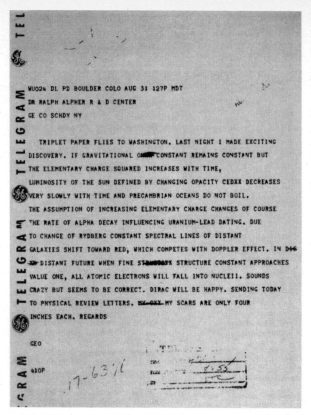

Figure 12.1 *Triplet Paper Flies to Washington! Gamow's telegram to his former student Ralph Alpher telling him of his idea that increasing electric charge can avoid making the oceans boil too recently in Earth's history.*[5]

measurements that were then possible it was found that the fine structure constant was the same then as now to within a few per cent:

$$\alpha(z = 1.95)/\alpha(z = 0) = 0.97 \pm 0.05$$

Soon afterwards, in 1967, Bahcall and Schmidt[6] observed a pair of oxygen emission lines that appear in the spectra of five galaxies which

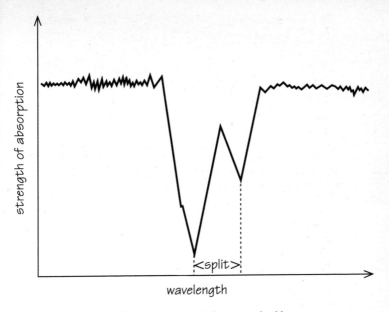

Figure 12.2 *Spectral lines in a typical atomic doublet system.*

emit radio waves, located at an average redshift of 0.2 (thus emitting their light about 2 billion years ago – around the time that the Oklo reactor was active on Earth) and produced a result consistent with no change in the fine structure constant that was ten times stronger still:

$$\alpha(z = 0.2)/\alpha(z = 0) = 1.001 \pm 0.002$$

These observations easily excluded the proposal by Gamow that the fine structure constant was increasing linearly with the age of the Universe. If that had been the case, the ratio $\alpha(z = 0.2)/\alpha(z = 0)$ should have been found to be about 0.8.

These ideas set the scene for astronomers to improve our knowledge of the constancy of particular constants of Nature as the improving sensitivity of telescopes and electronic detectors allowed observations to be made at higher and higher redshifts, reaching further and further back in time. The general strategy is to compare two atomic transitions in an astronomical site and here and now in the lab. For example, if they are doublets of elements like carbon, silicon or magnesium, which are commonly seen in gas clouds at high redshifts, then the wavelengths of two spectral lines, λ_1 and λ_2 say, will be separated by a distance that is proportional to α^2. The relative line shift is given by a formula

$$(\lambda_1 - \lambda_2)/(\lambda_1 + \lambda_2) \propto \alpha^2.$$

Now we need to measure the wavelengths λ_1 and λ_2 very accurately in the lab here, and far away by astronomical observations. By calculating the left-hand side of our formula to high accuracy in both cases we can divide our results to find that

$$[(\lambda_1 - \lambda_2)/(\lambda_1 + \lambda_2)]_{lab} / [(\lambda_1 - \lambda_2)/(\lambda_1 + \lambda_2)]_{ast} = \alpha_{lab}^2/\alpha_{ast}^2$$

We aim to discover if there is any significant deviation from 1 when we calculate the ratio on the left-hand side. If there is, it tells us that the fine structure constant has changed between the time the light left and the present. In order to be sure that there really is a significant deviation from 1, several things must be under very precise control. We need to be able to measure the wavelengths λ_1 and λ_2 to high accuracy in the lab. We also need to be sure that the observations are not being affected by extraneous noise, or biased by some subtle propensity of our instruments to gather certain sorts of evidence more readily than others.

Another approach is to compare[7] the redshifts of light emitted by molecules like carbon monoxide with that from atoms of hydrogen in the same cloud. In effect, one is measuring the redshift of the same cloud by two means and comparing them. This uses radio astronomy and allows us to compare the value of α here and now[8] with its value at the astronomical sources. When they are at redshifts 0.25 and 0.68

this leads to a limit on a possible shift, $\Delta\alpha$, in α between then and now of

$$\frac{\Delta\alpha}{\alpha} = \frac{\alpha(z) - \alpha(\text{now})}{\alpha(\text{now})} = (-1.0 \pm 1.7) \times 10^{-6}.$$

One of the challenges of this method is to make sure that the atomic and molecular observations are looking at atoms and molecules that are moving in the same way in the same cloud at their distant location.

A third method is to compare the redshift found from 21 cm radio observations of emissions from atoms with optical atomic transitions in the same cloud. The ratio of the frequencies of these signals enables us to compare the constancy of another combination of constants[9]

$$A \equiv \alpha^2 m_e / m_{pr}$$

where m_e is the electron mass and m_{pr} is the proton mass. Observation of a gas cloud at a redshift of $z = 1.8$ leads to a limit[10] on any change in the combination A of[11]

$$\Delta A/A = [A(z) - A(\text{now})]/A(\text{now}) = (0.7 \pm 1.1) \times 10^{-5}$$

The important thing to notice about these two results is that the measurement uncertainty is large enough to include the case of *no* variation:

$$\Delta\alpha/\alpha = 0 \text{ and } \Delta A/A = 0$$

It is important to stress that over the whole period from 1967 to 1999 when these observations were being made to increasing precision there was never any expectation that a non-zero variation of any traditional constant would be found. The observations were pursued as means of improving the limits on what the smallest allowed variations could be.

Their novelty was that they were so much more restrictive than any limits that could be obtained in the lab by direct experimental attack. Just watching the energy of an atom for a few years to see if it drifted cannot compete with the billions of years of history that astronomical observations can routinely monitor.

The fourth and newest method is the most powerful. Again, it looks for small changes in how atoms absorb light from distant quasars. Instead of looking at pairs of spectral lines in doublets of the same element, like silicon, it looks at the separation between lines caused by the absorption of quasar light by *different* chemical elements in clouds of dust lying between the quasar and us (see Figure 12.3).

There are a number of big advantages with this new method. It is possible to look at the separations between many absorption lines and build up a much more significant data set. Better still, it is possible to pick the pairs of lines whose separations are to be measured so as to maximise the sensitivity of the separations to little shifts in the value of α over time. But there is an unusual further advantage of this method. The wavelength separations that need to be extracted from the

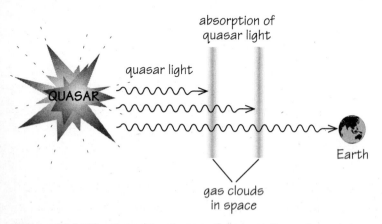

Figure 12.3 *The absorption of quasar light by different chemical elements in clouds of material lying between a far distant quasar and us.*

astronomical data and measured in the lab depend on α in distinctive ways. We can use large computer simulations[12] to discover what would happen to the positions of the lines if a tiny shift was made in the value of α. The shifts are very different for different pairs of lines. Increase α by one part in a million and some separations increase, some decrease, while some are almost unaffected. The whole collection of shifts defines a distinctive fingerprint of a shift in the value of α. Any spurious influence on the data, or messy turbulence at the site where the absorption is occurring out in the Universe, seeking to fool us into thinking that α is changing when it isn't, has got to mimic the entire fingerprint left on the wavelength separations by true α variation.

This method, called the many-multiplet (or MM) method by its inventors is far more sensitive than the other astronomical methods and allows much more of the information in the astronomical data to be used.[13] It has been applied by us to observations of 128 quasars, looking at separations between magnesium, iron, nickel, chromium, zinc and aluminium. When we began this work we expected that we would be able to use our new technique to place even stronger limits on the constancy of the fine structure constant. But we were in for a big surprise.

INCONSTANCY AMONG THE CONSTANTS?

'I feel like a fugitive from the law of averages.'

Bill Maudlin[14]

When we first developed the many-multiplet method we expected that it would lead simply to a further major improvement of the limits on any allowed change in the fine structure constant. It was an ideal method to exploit developments in extragalactic astronomy, big telescopes and new detector technology. Absorbing gas lying between us and distant quasars is a perfect laboratory for checking the constancy of constants because quasars are bright and easily accessible by telescopes over a wide range of redshifts. There are some constraints though. If you try to see objects at too high a redshift then the signals will be too faint to detect clearly. Also, unfortunately, some of the wavelengths of light that would be very interesting end up being redshifted en route to us so that they fall outside the window of wavelengths that can pass through the Earth's atmosphere to the ground.

The results gathered and analysed over two years by our team of John Webb, Mike Murphy, Victor Flambaum, Vladimir Dzuba, Chris Churchill, Michael Drinkwater, Jason Prochaska, Art Wolfe and me, with contributions of data by Wallace Sargent, proved to be unexpected and potentially far-reaching. If they are telling us what they appear to be telling us then, in the words of one commentator,[15] 'it'll be the most startling discovery of the past 50 years'.

We find a persistent and highly significant difference in the separation of spectral lines at high redshift compared to their separation when measured in the laboratory.[16] The complicated 'fingerprint' of shifts matches that predicted to occur if the value of the fine structure constant was *smaller* at the time when the absorption lines were formed by about seven parts in a million. If we combine all the results then the overall pattern of variation that results is shown in Figure 12.4.[18]

The first studies using the MM method in 1999 reported evidence for a variation in the value of the fine structure constant in

Figure 12.4 *The relative shift ($\Delta\alpha/\alpha$) in the value of the fine structure constant (in units of 10^{-5}) at different redshifts, and look-back times into the past measured in billions of years (Gyr). There is a significant negative shift between redshifts 1 and 3, indicating that the fine structure constant appears to have been smaller in the past by about seven parts in a million. (a) Shows all the astronomical objects observed. (b) Simplifies the data by collecting the data points in (a) into groups of ten observations.*

(a)

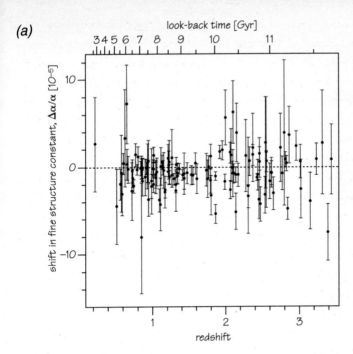

(b)

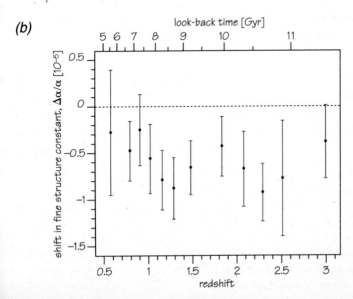

the past. Since then, the data has steadily increased and better analysis techniques have been employed. Remarkably, the same results are found from the whole collection of observations of 128 quasars. This is the largest direct observational assault on the question of whether the constants are the same now as they were thirteen billion years ago.

The first striking feature is that if we use the results to calculate what the fine structure was in the past we find a period in cosmic history where it appears to be slightly smaller than it is today. The magnitude of the dip in its value is very small, about seven parts in a million, and too small to have been found in any earlier investigations by observers using other methods, or detected in any laboratory experiment. It points to electricity and magnetism being slightly weaker in the past and atoms slightly bigger. If we take the observations of sources lying between redshifts of 0.5 and 3.5 as a whole, the observed shift is[19]

$$\Delta\alpha/\alpha = [\alpha(z) - \alpha(now)]/\alpha(now) = (-0.57 \pm 0.10) \times 10^{-5}$$

If one converts this into a rate of change of α with time it amounts to about

$$\{\text{rate of change of } \alpha\}/\{\text{current value of } \alpha\} = 5 \times 10^{-16} \text{ per year}$$

A first reaction to these dramatic results might be that they are claiming to find a variation that is much larger than is permitted by the evidence from the Oklo natural reactor studies. But on reflection they are not in direct conflict. Leaving aside all the uncertainties that go into finding the exact dependence of neutron capture rates in the Oklo reactor on the fine structure constant, the Oklo observations probe the fine structure constant's value only about 2 billion years ago (a redshift of about 0.1) whereas the quasar observations span the range from about 3 to 11 billion years ago. The two observations are only in conflict if you assume that the fine structure constant always changes at the same rate. But, as we shall see, there is no need to make any assumptions.

WHAT DO WE MAKE OF THAT?

'I hope I shall not shock the experimental physicists too much if I add that it is also a good rule not to put over-much confidence in the observational results that are put forward until they have been confirmed by theory.'

Arthur Eddington[20]

The evidence that the fine structure constant may have been different in the past is impressive but it is statistical in character. It is based upon the totality of astronomical observations of light absorption by many different chemical elements in nearly 128 different dust clouds. In the future more data will be added to the total and the question will be probed by better and better observations. Ideally, other astronomers should repeat our observations and use different instruments and different data analysis techniques to see if they get the same results.

Yet, desirable as they are, more observations and greater accuracy are not panaceas. In observational science one must be aware of different types of uncertainty and 'error'. First, there is uncertainty introduced by the limiting accuracy of the measurement process. If your height is being measured to the nearest centimetre and is quoted as being 1.85 metres, it could actually lie anywhere between 1.845 and 1.855 metres. This type of uncertainty is usually well understood and can gradually be reduced by improving technology (use a more finely graduated ruler). Second, there is a subtler form of uncertainty, usually called 'systematic error' or 'bias', which skews the data-gathering process so that you unwittingly gather some sorts of evidence more easily than others. More serious still, it may ensure that you are not observing what you thought you were observing.[21]

All forms of experimental science are challenged by these subtle biases. In down-to-earth laboratory subjects it is usual to repeat experiments in several ways, changing certain aspects of the experimental set-up each time, so as to exclude many types of bias. But in astronomy there is a bit of a problem. There is only one Universe. We are able to observe it but we can't experiment with it. In place of experiment we look for correlations between different properties of objects: do all the clouds with particular redshifts have smaller spectral shifts between certain absorption lines, for instance? One might be aware of a bias and yet be unable to correct completely for its influence, as in the case of creating a big catalogue of galaxies where one is aware of the simple fact that bright galaxies are easier to see than faint ones. But the real problem is the bias that you *don't* know about. The data used to study the possible variation in the fine structure constant has been subjected to a vast amount of test and scrutiny to evaluate the effects of every imaginable bias. So far, only one significant influence has been found and accounting for it actually makes the deduced variations *bigger*.[22]

The reaction of most physicists or chemists to the idea that the fine structure constant might be changing by a tiny amount over billions of years is generally one of horror and outright disbelief. The whole of chemistry is founded on the belief in theories which assume that it is absolutely constant. However, a change of a few parts in a million over 10 billion years would have no discernible effect upon any terrestrial physics or chemistry experiment. To see this more clearly it is time to ask what exactly are the best direct experiment limits that we have on the change in the fine structure constant.

Most direct tests of the constancy of the fine structure constant take an atom and monitor it for a given length of time as accurately as the measuring set-up will allow, typically to a few parts in a billion. This amounts to comparing different atomic clocks. This monitoring cannot be carried out for very long because of the need to keep other things constant, and the best results have come from a run of 140 days.[23] Assuming that the ratio of the electron and proton masses does not change, experimenters find that the stability of the value of an energy transition between hydrogen and mercury means that if the fine structure constant is changing then its rate of change must be less than 10^{-14} per year. This result sounds very strong. It allows the constant to change by only about one part in 10,000 over the whole age of the Universe but the astronomical observations are recording a variation that is about 100 times smaller still. This gap between lab and outer space also illustrates the huge gain in sensitivity that the astronomical observations offer over the direct lab experiments. They may not be making measurements of the fine structure constant at the technological limit of sensitivity but they are looking so far back into the past – 13 billion years instead of 140 days – that they provide far more sensitive limits.[24] The Universe has to be billions of years old in order that stars have enough time to create the biological elements needed for living complexity to exist within it. If those complicated pieces of chemistry happen to be astrophysicists then it is a nice by-product of the Universe's great age that such sensitive probes of Nature's constancy will be available to them.

So it seems that we cannot use terrestrial experiments to double check the apparent changeability of the fine structure constant – we just don't have instruments sensitive enough to pick up a variation at the level seen in the astronomical data. At the moment the best chance of an independent confirmation from a completely different direction would seem to lie with some other astronomical probe. Oklo tells us that we should not expect to find a similar rate of variation more recently, two billion years in the past, but perhaps such a variation could exist and have observable effects in the very early stages of the Universe's history. The quasars reach back through 80 per cent of the Universe's history but we can see back a lot further than that by probing the microwaves that were left over from the beginning of the Universe's expansion. This is usually called the cosmic microwave background radiation and it ceased interacting with matter when the

Universe was only a few million years old. Whereas the quasars we were observing had redshifts up to a value of 3.5, the microwave radiation was effectively emitted at a redshift of 1100. Its structure is giving us a snapshot of the Universe's shape and evenness when it was only 300,000 years old (see Figure 12.5).

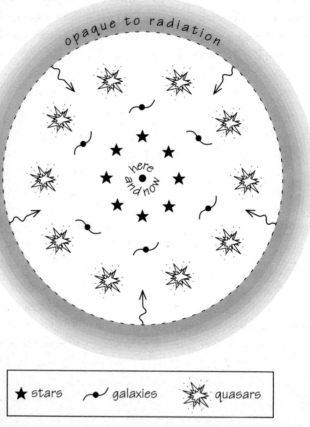

★ stars   galaxies   quasars

Figure 12.5 *Looking out in space (and backwards in time) we reach the epoch when quasars form. Further out, we reach the surface where the background radiation was opaque and all atoms were dismembered by the heat radiation. This occurred when the Universe was only about 300,000 years old and about 1000 times smaller than today.*

In recent years astronomers have made newspaper headlines all over the world by mapping this radiation in exquisite detail with receivers carried on balloons and satellites. We know that the radiation has the spectrum of pure heat radiation to very high precision and its temperature is the same in different directions on the sky to an accuracy of about one part in 100,000. The detailed maps that are being constructed of the statistics of its temperature variations on the sky hold the secrets of what the galaxies and clusters were like in their extreme youth, when they were little more than embryonic islands of material slightly more dense than the rest of the Universe around them.

Unfortunately, there does not seem to be a clean and simple diagnostic of the value of the fine structure constant when the microwaves were transmitted to us. However, motivated by our results from quasars, several teams of cosmologists have carried out a complicated reconstruction of what the statistical pattern of fluctuations should look like on the sky if α had a different value at a redshift of 1100. They have to use the most reasonable theories of how fluctuations that will grow into galaxies affect the microwave temperature patterns on the sky. Interestingly, they claim that the most recent data is slightly better understood if there is a smaller value of the fine structure constant at this high redshift.[25] The magnitude of change required is huge – ten per cent[26] – and would require a steady fall in the value of α as we went backwards in time from the quasar epoch to the last scattering of the microwave radiation. This is not a very compelling piece of evidence given the large number of variants on the whole picture for the formation of galaxies. There are too many other small effects on the temperature pattern, all quite reasonable, which produce an overall effect like that attributable to a smaller value of the fine structure constant. Without more information about what to look for this does not look like a promising route to discovering the value of the fine structure constant in the past. But things may change. During 2002 NASA's Microwave Anisotropy Probe (MAP) satellite will send back new all-sky maps of the microwave background radiation and its pattern of variations. The unprecedented accuracy expected from this instrument may allow new conclusions to be drawn early in 2003.

OUR PLACE IN HISTORY

> 'That rooster was like an impatient person. Like someone who lived in the city, someone who always seemed to have too much to do, but never did anything but attend to his own haste. Life wasn't like that in the village: here everything moved as slowly as life itself. Why should people hurry when the plants that nourished them grew so slowly?'
>
> Henning Mankell[27]

If the constants of Nature are slowly changing then we could be on a one-way slide to extinction. We have learnt that our existence exploits many peculiar coincidences between the values of different constants of Nature, and that the observed values of the constants fall within some very narrow windows of opportunity for the existence of life. If the values of these constants are actually shifting, what could happen? Might they not slip out of the range that allows life to exist? Are there just particular epochs in cosmic history when the constants are right for life?

There are two situations where it is possible to examine the changes in traditional constants in some detail. For only when the fine structure 'constant', α, or Newton's gravitational 'constant', G, are changing do we have a full theory which includes the effects of the changes. These theories are generalisations[28] of the famous general theory of relativity created by Einstein in 1915. They allow us to extend our picture of how an expanding universe will behave to include variations of these constants. If we know something about the magnitude of a variation at one epoch we can use the theory to calculate what should be seen at other times. In this way the hypothesis that the constants are varying becomes much more vulnerable to observational attack.

If constants like G and α do *not* vary in time, then the standard history of our Universe has a simple broad-brush appearance. During the first 300,000 years the dominant energy in the Universe is radiation and the temperature is greater than 3000 degrees and too hot for any atoms or molecules to exist. The Universe is a huge soup of electrons, photons of light and nuclei.

We call this the 'radiation era' of the universe. But about 300,000 years there is a big change. The energy of matter catches up with and overtakes that of radiation. The expansion rate of the Universe is now primarily dictated by the density of atomic nuclei of hydrogen and helium. Soon the temperature falls off enough for the first simple atoms and molecules to form. Over the next 13 billion years a succession of more complicated structures are formed: galaxies, stars, planets and, eventually, people. This is called the 'matter era' of the Universe's history. But the matter era might not continue right up to the present day. If the Universe is expanding fast enough then, eventually, the matter will not matter, and the expansion just runs away from the decelerating clutches of gravity, like a rocket launched at more than the escape speed from Earth. When this happens we say the Universe is 'curvature-dominated' because the rapid expansion creates a negative curvature to astronomical space, just like that near the seat of a riding saddle.

There are three trajectories for an expanding universe to follow (see p. 184). The 'closed' universe expands too slowly to overcome the decelerating effects of gravity and eventually it collapses back to high density. The 'open' universe has lots more expansion energy than gravitational deceleration and the expansion runs away forever. The in-between world, that is often called the 'flat' or 'critical' universe, has a perfect balance between expansion energy and gravity and keeps on expanding for ever. Our Universe is tantalisingly close to this critical or 'flat' state today.

Another possibility is that the vacuum energy of the universe can eventually come to dominate the effects of the ordinary matter and cause the expansion of the universe to begin accelerating. Remarkably, current astronomical observations show that our Universe may have begun to accelerate quite recently, when the Universe was about three-quarters of its present age. Moreover, these observations imply that the expansion of our Universe has not become curvature-dominated. The overall pattern of the expansion history since it was about a second old is shown in Figure 12.6. The observations are telling us that about 70 per cent of the energy in the Universe is now in the vacuum form that acts to accelerate the expansion whilst almost all the rest is in the form of matter.

What happens to this story if the fine structure constant changes? The expansion is virtually unaffected by the variations in the fine structure constant if they are as small as observations suggest – a million

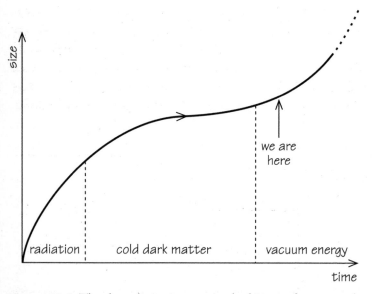

Figure 12.6 *The three distinctive eras in the history of an expanding universe like ours, which appears to have about 70 per cent of its present energy in an unknown form of vacuum energy that acts to accelerate the expansion. The expansion of such a universe has three distinct eras, dominated by radiation, cold dark matter, and vacuum energy.*

times slower than the Universe is expanding – but the expansion dramatically affects how the fine structure 'constant' changes.

Håvard Sandvik, João Magueijo and I investigated what would happen to the fine structure constant over billions of years of cosmic history. The conclusions were rather striking but appealingly simple. During the radiation era there is no significant change at all. But once the matter era begins, when the Universe is about 300,000 years old, the fine structure constant starts to *increase* in value very slowly.[29] When a curvature era begins, or the vacuum energy begins to accelerate the Universe, that increase stops. This characteristic history is shown in Figure 12.7 for a universe with matter, radiation and vacuum energy values equal to those we observe in our Universe today.

This is intriguing. It paints a picture that fits all the evidence rather well. Our Universe began accelerating at a redshift of about 0.5 and so there will be no significant variation of the fine structure at the time of the Oklo reactor. Over the interval of redshifts corresponding to the quasar observations the variations can be of the form that is seen and α is predicted to be smaller in the past: just what we see. If we keep going back to the redshift around 1100 where the microwave radiation

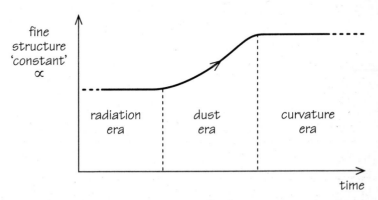

Figure 12.7 *The expected change in the fine structure 'constant' in a universe like ours: the 'constant' stops changing when the universe begins to accelerate and only changes very slowly during the period of cold matter domination.*

starts flying freely towards us we predict that variation in α should be much smaller than the sensitivity of the present observations.

If these variations really are taking place as the Universe expands then they have consequences for the evolution of life. We know that if the fine structure 'constant' grows too large then atoms and molecules will be unable to exist and no stars will be able to form because their centres will be too cool to initiate self-sustaining nuclear reactions.

It is therefore crucial that the dust era of cosmic history during which the fine structure constant increases does not last too long. Without the vacuum energy or the curvature to stop the steady increase in the fine structure constant's value there would come a time when no life is possible. The Universe would cease to be habitable by atom-based forms of life who relied upon stars for energy.

Something similar happens if there can be variations in the strength of gravity, represented by the Newtonian 'constant' G. During the radiation era it tends to stay constant but when the matter era begins it starts to fall in value until the curvature era begins. If the Universe never experiences a curvature era then gravity just keeps getting weaker and weaker and it becomes harder and harder for planets and stars to exist. The behaviour is shown in Figure 12.8.

This overall evolution is very intriguing. It shows that even when the constants are allowed to vary they are only able to exploit that freedom to vary when the Universe is in the matter era. If they *are* varying then we are observing the Universe during a niche of history when these constants have values that allow atoms, stars and planets to exist.

It has always been something of a mystery why our Universe is so close to the critical state of expansion today and why the vacuum energy is so fantastically small. We know that if we were too far from the critical expansion rate then life would have been far less likely to have evolved on Earth, and would probably be impossible anywhere else in the Universe as well. If universes are too curvature-dominated then the expansion goes so fast that islands of material cannot overcome the effect of the expansion and contract to form galaxies and

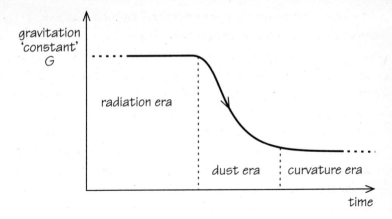

Figure 12.8 *The typical behaviour of a changing gravitation 'constant' over cosmic history in cosmological theories which allow such changes to occur. The strength of gravity only changes significantly during the era when cold dark matter ('dust') dominates the universe and is switched off by the effects of radiation or spatial curvature dominating the expansion of the universe.*

stars. On the other hand, if the Universe expands too slowly it soon collapses back to a Big Crunch. Dense islands of material form too quickly and fall into large black holes before stars and biochemistry have a chance to form (see Figure 9.2).

Likewise, with the vacuum energy. If it were ten times bigger it would have started accelerating the Universe's expansion so early on in its history that galaxies and stars would not have been able to separate out from the overall expansion.

Both these arguments show us that we should not be surprised to find that the deviations from the critical expansion rate, or from having zero vacuum energy in the Universe, are not large. We would not be here if they were. But the possibility of varying constants provides us with a possible reason why the Universe could not be observed by us to be exactly critical and to have no vacuum energy.[30] The vacuum energy and the curvature are the brake-pads of the Universe that turn off variations in the constants of Nature. They stop the constants changing. If their variations are not curtailed then they will reach values that prevent the existence of atoms, nuclei, planets and stars. The Universe will eventually become lifeless, unable to contain the building blocks of complexity. Then life, like all good things, must come to an end.

chapter thirteen

Other Worlds and Big Questions

'O World of many worlds, O life of lives,
What centre hast thou? Where am I?'

Wilfred Owen[1]

MULTIVERSES

'The apparent uniqueness of the Universe primarily depends upon the fact that we can conceive of so many alternatives to it.'

Charles Pantin[2]

Our excursions along the new pathways that have been opened up by our attempts to understand and explain the values of the constants of Nature raise a host of big questions about the nature of things. We have seen that cosmologists actively contemplate the nature of 'other worlds' in which the constants of Nature take different values to our own. It appears that very small changes to many of our constants would make life impossible. This raises the deeper question of whether these other worlds 'exist' in any sense and, if so, what it is that makes them different from the world we see and know. It also provides an alternative to the old argument that the apparent fine tuning of the world to possess all those properties required for life is evidence for some form of special design. For, if all possible alternatives exist, we must necessarily find ourselves inhabiting one of those that permits life to exist. And we could go still further and hazard a guess that we might expect to find ourselves in the most probable sort of life-supporting universe.[3] The first person who seems to have articulated this many-worlds approach was the Cambridge biologist Charles Pantin, who tried to find a more appealing context for thinking about special properties of the Universe's structure, constants and laws by introducing the notion of an ensemble of many worlds, each with a different suite of physical properties:

> 'If we could know that our own Universe was only one of an indefinite number with varying properties we could perhaps invoke a solution analogous to the principle of natural selection; that only in certain universes, which happen to include ours, are the conditions suitable for the existence of life, and unless that condition is fulfilled there will be no observers to note the fact.'[4]

One of the difficulties of even conceiving of such a multiverse of all possible universes is that there are so many things that could be different. From our study of mathematics we know that there exist different logics to the one that we use in practice, in which statements are either true or false. Likewise, there are different mathematical structures; different possible laws of Nature; different values for the constants of Nature; different numbers of dimensions of space or of time; different starting conditions for the universe; and different random outcomes to complex sequences of events. On the face of it the collection of all possible worlds would have to include, at the very least, all the possible permutations and combinations of these different things. Gaining an understanding of this cornucopia is a tall order.

We have already seen what might happen if some of the other

possible worlds were to be actualised, worlds with more dimensions or other values of crucial constants. However, we don't know whether these different worlds really are possible worlds. It is all very well to contemplate changes in the values of the constants of Nature and the quantities that define the shape and size of the universe. But are they really allowed alternative universes or are they no more possible than square circles? It could be that the all-encompassing Theory of Everything is very restrictive when it comes to giving planning permission for other universes. The fact that we can conceive of so many alternative universes, defined by other values of the constants of Nature, may be simply a reflection of our ignorance about the strait-jacket of logical consistency that a Theory of Everything demands.

When it comes to contemplating other universes there are two ways in which the problem can be approached. There is the conservative approach that produces alternative worlds by making little changes to the properties of our world – small shifts in the values of some of the constants of Nature, slightly different properties of the astronomical Universe, perhaps, but no changes to the laws of Nature themselves. Typically, these studies show that if the 'little changes' are too large there are adverse consequences for the existence of life as we know it. Our sort of life could still exist if there was a one part in a hundred billion change in the value of the fine structure constant, we think,[5] but not if it was a one part in ten change. By contrast, the radical approach thinks about big changes, where things like the laws, the underlying mathematical logic, or the number of dimensions of space and time can be changed. It has to conceive of completely new types of 'life' which could exist in entirely different environments.[6] This provokes a closer examination of what is meant by 'life'. Typically, it is reduced to some bare essentials, like the ability to process and store information (if you are a computer scientist), the facility to evolve by natural selection (if you are a biologist), or simply non-equilibrium energy flow (if you are a chemist).

As an example of the radical approach, consider the search for 'life' in mathematical formalisms that I once proposed.[7] We consider the hierarchy of all possible mathematical structures, starting with simple finite collections of points related by rules, then geometries, then counting systems like the arithmetic of whole numbers, then fractions, then decimals, then complex structures and groups and so on, onwards and upwards forever, on an ascending staircase of complexity. Now we ask which of these structures can fully describe conscious beings. For if we were to take the axioms of one of these systems of logic, and then gradually work out all the truths that can be deduced from them, using the prescribed rules of deduction, we would see a great web of logical truths stretching out before us. If that web of truth eventually leads to structures that completely describe what we call 'consciousness' then they could be said to 'be alive' in some sense. The question is: in what sense?

Another way to look at this is to think about the creation of a computer model, or simulation, of the process by which stars and planets form. This is something that astronomers work hard to do. Star formation is too complicated to understand in full detail just by using pencil and paper and direct human calculation. Fast computer solution of the equations that govern it is needed. Let us imagine that in the far future these simulations have become extremely accurate. They describe how stars form and produce descriptions of planets that match very closely with those that we see. We judge this problem to be 'solved'. An enthusiastic biochemist suggests that we go a bit further and feed into the computer lots of information about biochemistry and geology so that we can follow the computer's predictions about the early chemical evolution of a planet and its atmosphere. When this is done the results are very interesting. The computer describes the formation of self-replicating molecules that start to compete with each other and do complicated things on the young planet's surface. Helices of DNA appear and start to form the bases of genetic replicators. Selection begins to make an impact and the best adapted replicators multiply and improve very quickly, spreading their blueprints all over the habitable surface. The computer programme is run for longer and longer. Eventually, some of the structures in the programme seem to be signalling to one another and storing information. They have developed a simple code and what we would call an arithmetic, that is based upon the symmetry (eight-sided) that the biggest replicators possess. The programmers are fascinated by this behaviour, never suspecting that it could emerge from their original programming. The behaviour of the replicators is like a code and at first it is not too difficult to break. The patterns in the computer output are developing a simple logic for communication. A video display of the output makes it all seem like a natural history film about the evolution of life.

This little fantasy shows how it is conceivable that behaviour we might judge to be conscious could emerge from a computer simulation. But if we ask where this conscious behaviour 'is' we seem to be pushed towards saying that it lives in the program. It is part of the software running in the machine. It consists of a collection of very complex deductions ('theorems') that follow from the starting rules that define the logic of the programming. This life 'exists' in the mathematical formalism.

These examples seek to capture one aspect of life as a computer programme. They are suspiciously powerful because they lead to the conclusion that if 'life', suitably defined, *can* exist in a mathematical formalism then it *does* exist in the fullest sense.[8] This is not unlike Anselm's famous ontological argument for the necessary existence of God.

The problem with such computer ontological arguments that allow life to be placed within mathematical formalisms is that they equate mathematical existence with physical existence. Physical existence is something that we have some experience of. We probably can't define it but, like many things that we have difficulty defining, we know it when we see it. Mathematical existence is a far weaker thing, but much easier to define. Mathematical existence just means logical self-consistency: this is all that is needed for a mathematical statement to be 'true'. Thus right-angled triangles 'exist' in the system of Euclid's geometry. Square circles don't.

A true mathematical statement doesn't have to be interesting; it doesn't have to be short; it doesn't have to be new. It just must not lead to a logical contradiction with the logical rules being used.[9] These mathematical universes can be imaginary in many senses. A few, like the mathematician Godfrey Hardy (1877–1947), have thought that some of them are more appealing than the actual:

> '"Imaginary" universes are so much more beautiful than this stupidly constructed "real" one; [but] most of the finest products of an applied mathematician's fancy must be rejected, as soon as they have been created, for the brutal and sufficient reason that they do not fit the facts.'[10]

A possible objection to obtaining life-supporting worlds as outputs from some great computer code is that there seem to be so many more mathematical formalisms that don't lead to life than those that do. But that's all right. Our anthropic argument has taught us that we must find ourselves in one of those that does support life. However, there is a more subtle problem. There are also an infinite number of universes that possess the lawful ordered structure that we see around us up until the present moment, but which will behave in a completely different or lawless fashion from now onwards. It therefore seems far more likely that we will live in a universe where our belief that the Sun will rise tomorrow fails.[11] If there are so many more possible worlds in which the Sun does not rise tomorrow but in which everything is just the same as in our life-supporting world up until sunrise tomorrow, what should we conclude if the Sun does rise tomorrow?

This is not the paradox that it first appears. It requires some way of assessing the likelihood of the different histories. The most appropriate method may not just be to count them. The histories that are orderly up to a point and then diverge into chaos require a specification that makes them less likely in the space of all possibilities than the ones that continue in the same state of life-supporting orderliness.

These other worlds are rather Platonic. Their existence does not smack of what we like to think of as 'real' existence. It's virtual rather than real. Somehow life in a mathematical formalism or within a computer program isn't really living. But maybe all conscious information-processors in these formalisms suffer from similar delusions of grandeur and uniqueness. But let's suppose they are right and move on to some more concrete ensembles of other worlds.

THE GREAT UNIVERSAL CATALOGUE

'The universe is merely a fleeting idea in God's mind – a pretty uncomfortable thought, particularly if you've just made a down payment on a house.'

Woody Allen[12]

Cosmologists have considered ways in which some of the ensembles of other worlds might arise. Generally, they spring from the conservative approach to creating other worlds that we introduced above. We consider a small number of changes to the Universe we know, leaving the laws unchanged but altering the values of its constants or its dimensions. We have already seen the case of the inflationary universe in its chaotic and eternal editions. Different large regions of our single Universe, which may be infinite in size, can find themselves with different average densities, different expansion rates, or even different numbers of large dimensions of space and different forces of Nature, as a result of the randomness intrinsic to the processes that initiate inflation. Inflation may have begun and ended at different times in different places. The result would be a universe containing different regions where conditions were very different and the values of some of the defining constants that are essential for life could take different values (see Figure I3.Ia, b).

These regions are most likely to be very large, much larger than our visible Universe. Inflation very easily expands small regions to large ones and so the boundary of our domain is most likely to be vastly larger than what we can see of the Universe. But one day our descendants might see the evidence of one of these regions where things are different coming over the astronomical horizon, annihilating distant matter, distorting the expansion of the Universe, and swallowing up stars and galaxies.

If the eternal edition of inflation is contemplated then the ensemble of possibilities expands even more, and we have to see ourselves as a local fluctuation in a never-ending process which explores all of the permutations of cosmic conditions, constants and dimensions that are open to it. Only in some will life be possible.

An interesting feature of these inflationary ensembles is that they do not ask us to believe in some multiverse of other worlds of dubious status. They are not parallel worlds or imaginary worlds, and may not even be merely hypothetical worlds. What counts as a 'world' is just a very large region of our one and only Universe. And if our Universe is infinite in extent then the number of alternatives that inflation can generate may be infinite as well. If it exhausts all the logical possibilities for variation that are available to it then any possibility that can exist will exist somewhere, not just once, but infinitely often. One thing we can say with certainty about this idea is that if it is true it cannot be original.[13]

There are other more mundane ways of producing large numbers of different possibilities within our single Universe. Nature creates complexity by breaking the symmetries of the laws of Nature in the outcomes of those laws. Thus at this moment you are located at a particular place in the Universe even though the laws of gravity and electromagnetism, of which you are a complicated outcome, have no preference for any places in the Universe. As the Universe expands and

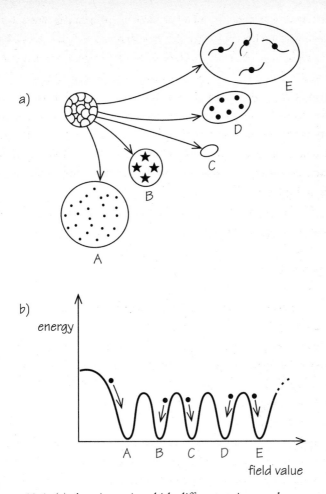

Figure 13.1 *(a) A universe in which different regions undergo a different amount of inflation, giving rise to different conditions. (b) The lowest energy state for the material in the universe at the end of inflation may not be unique. The universe may end up in a different minimum in different places. As a result the number and strength of the forces of Nature will be different in different places depending upon which minimum was reached for the matter in that locale.*

cools through its earliest phases there are a number of occasions when symmetries break. In some places it will break in one way, elsewhere in another. These random outcomes can have far-reaching consequences for the evolution of life in the future. A typical example of vital symmetry breaking is that which gives rise to the balance between matter and antimatter in the early Universe. As a result, the imbalance between matter and radiation that is needed to prevent everything being annihilated to radiation later will vary from place to place. If it were to happen before inflation occurs then a region that had a preponderance of matter would inflate and become a huge region containing our visible Universe. If it occurred after inflation then our visible portion of the Universe could contain regions with different balances of matter and antimatter. Again, we have a vehicle for creating large regions within a single Universe where some features which are critical for the existence of life can vary significantly from place to place.

The quantum description of the Universe teaches us that all the substantive things that we see and experience as particles or aggregates of matter have a wavelike quality. That quality expresses the probability that we will observe them to have certain properties. One of the interesting discoveries made by physicists wrestling with the problem of creating a quantum description of the entire Universe is that the starting conditions for the Universe seem to play a crucial role in the transition from wavy to substantive properties.

We are used to the idea that the wavy indefinite nature of particles of matter is something that occurs in the realm of the very small. When things become large this quantum waviness becomes small and negligible. We might have to worry about it when we do atomic physics but we don't need to when we are driving a car. However, it appears that this substantive quality of our experience — that there is definite non-quantum behaviour of things — is not guaranteed in all universes that become big and old. Special starting conditions seem to be needed for this to happen. In many worlds our familiar qualities like position, energy, momentum and time will never emerge in a well-defined way and nor, we suspect, will the type of complex organisation we call life.

The modern search for a Theory of Everything offers scope for other worlds as well. It is often imagined that the ultimate Theory of Everything will specify all the constants of Nature, but this now looks far less likely. It appears that only a fraction of the constants of Nature will be fixed absolutely by the inflexible internal logic of the theory while the others will be free to take different values that are chosen by a haphazard, symmetry-breaking process. As we saw in Chapter 8, faced with this openness, we have to turn to anthropic selection to explain why we see the values in the narrow life-supporting ranges that we do.

So far we have been content to create ensembles of other worlds by tinkering with parts of our world and exploiting its natural propensity to make things fall out differently from place to place. It is time to be more speculative and to consider some of the ways in which the constants of Nature could be made to change and fill out the collection of all possibilities by stepping outside the constraints of mainstream theories of physics into the realm of more speculative possibilities.

WORLDS WITHOUT END

'Universes that drift like bubbles in the foam upon the River of Time.'

Arthur C. Clarke[14]

Before the self-reproducing character of the eternal inflationary Universe[15] came to light it was suggested that it might be possible to initiate inflation in one part of the Universe by arranging particular high-energy collisions between elementary particles.[16] The eternal inflationary scenario is actually based upon the expectation that no arranging is really necessary. The Universe brings about the continual bouts of inflation without intelligent help or unintelligent mishap.

Now what if the Universe is eternally reinventing itself in bouts of inflation? Perhaps there have been super-advanced civilisations in regions that inflated in the past who *did* know how to initiate inflation and tailor its consequences. If so, they might be able to tune the outcomes of inflation to be advantageous for the continued existence of life. The British cosmologist Edward Harrison has speculated[17] that we could imagine such enlightened beings would choose to make the properties of the next edition of the Universe better suited for life than those within the one in which they had themselves evolved. If this tuning process continued over many generations of eternal inflation then we would expect the life-supporting 'coincidences' between the values of the controllable constants of Nature to become increasingly fine-tuned. Perhaps, Harrison suggests, this is why we find them so to be. Appealing as this intelligent design of universes might appear to be, it is not clear how it gets started. If universes begin with constants far from the values that allow complexity to develop, they will never develop the conscious beings needed to fine tune the constants. They will have to rely on random fluctuations to deliver a universe able to evolve beings intelligent enough to fine tune its constants.

Another interesting scheme which also sees the constants of Nature evolving under some external influence has been suggested by the American physicist Lee Smolin.[18] He suggests that every time a black hole forms in the Universe there is scope for a new parallel universe to emerge out of the mysterious singularity that develops at its centre. Everything that is captured by a black hole ends by falling inexorably into this singularity at its centre. Instead of disappearing into timeless oblivion the disappearing material is reborn as a new expanding universe with the values of its constants of Nature slightly shifted in a random way.[19]

In the long run this scenario leads to definite expectations. If the collapse of matter into black holes always spawns new universes then the more black holes a universe can produce the more offspring it will have to carry forward information about its own 'genetic code' – the values of its defining constants of Nature. Eventually, it is argued, we would expect to find ourselves living in a universe in which the constants have evolved towards a suite of values which maximise the production of black holes. Any little change in the values of the constants we observe should therefore make it *harder* to produce black holes.

This is just one of the conclusions that could be drawn from this scenario, though. From our anthropic considerations we can see that it might turn out to be the case that universes with constants taking values which maximise the production of black holes are unable to contain living observers at all. An application of the Anthropic Principle is therefore essential. We can predict only that we should find ourselves in a universe with constants whose values maximise the production of black holes, given that living observers must also be possible. And that may be a very different sort of universe.

Another long-term possibility is that there are no local maxima for the production of black holes when the constants are changed in value. There may be a direction of change for some constants which allow the black hole production to keep on getting bigger and bigger forever. Again, we can then say rather little about the ultimate values of the constants of Nature.[20]

This suggests another way in which an ensemble of other worlds with different constants can be generated from our Universe. If a universe contains sufficient matter to contract back upon itself and experience a Big Crunch in the future then it is a mystery what occurs at the Crunch. Physically, it is not very different from the centre of a black hole. Maybe the universe, along with space and time and the laws of Nature, just comes to an end and nothing follows. But cosmologists have always been tempted to think that the collapsing universe might 'bounce', phoenix-like, back into a state of expansion. If so, the natural conclusion to draw is that the universe will go on oscillating between states of expansion and contraction forever, as in Figure 13.2. The big question is what, if anything, changes when a bounce occurs? Is the sheet wiped clean or does some information about the old cycle get carried into the new one?

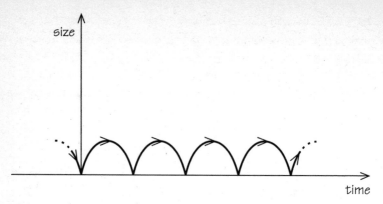

Figure 13.2 *An oscillating universe in which collapse to a future Big Crunch is succeeded by re-expansion into a new cycle, and so on for ever.*

It could be, as John Wheeler originally suggested, that there is a reshuffling of the values of the constants of Nature each time a bounce occurs.[21] This will create a never-ending sequence of expanding and contracting universes in which the constants are different. Only in those cycles in which the 'deal' of the constants comes out in a permutation that allows life to exist could we exist. Unfortunately, we have no idea how to link the values of constants in one cycle to those in the next. When it comes to the properties of the universe as a whole, there is one big factor that will play a dominant role. If the constants are changed into a permutation that does not allow the universe to collapse again to a Big Crunch the game will be over and the universe will be left stuck with a hand of constants that are never again re-dealt. Clearly, this is the most likely state for the universe to find itself in. If there have been an infinite number of oscillations of the universe in the past and there is any chance of a permutation being found which ends the oscillations then, eventually, that permutation will be realised and the oscillations will end.[22]

There is a favourite piece of continuity that cosmologists like to impose on the cycle-to-cycle evolution. It is the second law of thermodynamics, the principle that disorder ('entropy') never decreases with the passage of time. If this is adhered to from cycle to cycle and energy is conserved[23] then it causes the cycles to grow steadily in size (Figure 13.3).[24]

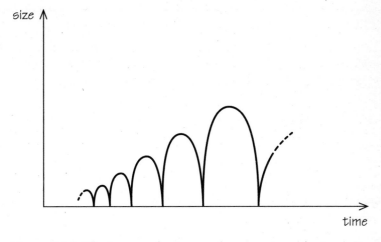

Figure 13.3 *The increase of entropy makes successive cycles increase in size if energy is conserved.*

This is rather interesting because in the long run the universe will push closer and closer to the state of critical expansion that inflation was invoked to explain. Yet there is one further twist to the story. Mariusz Dąbrowski and I[25] showed that if there is a cosmic vacuum energy acting to accelerate the expansion of the universe, as current observations suggest, then it will always bring the sequence of oscillations to an end and leave the universe on an accelerating ever-expanding trajectory into the future (Figure 13.4).

The end result is always to leave the universe trapped with its last deal of constants expanding in a state with a fine balance between the vacuum energy stress and all the other forms of matter in the universe, a little like our own Universe, in fact.

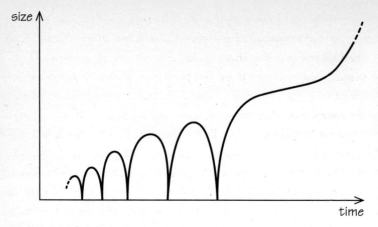

size

time

Figure 13.4 *If there is a small positive cosmological constant then eventually the cycles must come to an end, leaving the universe to expand for ever, accelerated by the influence of the cosmological constant.*

JOURNEY'S END

'Until the Scientific Revolution of the seventeenth century, meaning flowed from ourselves into the world; afterward, meaning flowed from the world to us.'

Chet Raymo[26]

Our look at the constants of Nature began with the mundane but has led us to the frontiers of our Universe, and even beyond, into a multiverse of other worlds whose existence we see only dimly reflected in that of our own. The search for standards that were humanly convenient and parochial led to the discovery of standards that were superhuman and universal. Our uncovering of the patterns by which Nature works and the rules by which it changes led us to the mysterious numbers that define the fabric of all that is. The constants of Nature give our Universe its feel and its existence. Without them, the forces of Nature would have no strengths; the elementary particles of matter no masses; the Universe no size. The constants of Nature are the ultimate bulwark against unbridled relativism. They define the fabric of the Universe in a way that can side-step the prejudices of a human-centred view of things. If we were to make contact with an intelligence elsewhere in the Universe we would look first to the constants of Nature for common ground. We would talk first about those things that the constants of Nature define. The probes that we have dispatched into outer space carrying information about ourselves and our location in the Universe pick on the wavelengths of light that define the hydrogen atom to tell where we are and what we know. The constants of Nature are potentially the greatest shared physical experience of intelligent beings everywhere in the Universe. Yet, as we have followed the highways and by-ways of the quest to unravel their meaning and significance, we have come full circle. Their architects saw them as a means of lifting our understanding of the Universe clear from the anthropomorphisms of human construction to reveal the otherness of a Universe not designed for our convenience. But these universal constants, created by the coming together of relativistic and quantum realities, have turned out to underwrite our very existence in ways that are at once mysterious and marvellous. For it is their values, measured with ever greater precision in our laboratories, but still unexplained by our theories, that make the Universe a habitable place for minds of any sort. And it is through their values that the uniqueness of our Universe is impressed upon us by the ease with which we can think of less satisfactory alternatives.

Will we ever explain the values of all the constants of Nature? So far, the answer is unclear, but in suggestive ways. Our deepest theories of the forces and patterns of Nature suggest that a Theory of Everything will have an openness about it. Not everything will be pinned down by the dead hand of logical consistency. There are some constants that have the freedom to be different; that are chosen at random; and that could render the Universe devoid of life and light forever if they fell out wrong rather than right.

And what of the very nature of these constants? Are they truly constant, the same yesterday, today, and forever, or are they merely ebbing and flowing slowly with the tides of time? As we look with our finest instruments we have begun to see the first tell-tale hints of a change in one of our most revered constants of Nature over billions of years of cosmic history. What does this mean for our understanding of the jigsaw of pieces that we assemble into our picture of the Universe? Will the constants change and destroy the coincidences between their values in the future and leave the tree of life leafless and lifeless in the far-distant future? Are our constants linked to the overall rate of expansion of our Universe or are they truly constant, an insulation from the evolution of complexity, life and the whirl of gravitating stars and galaxies around us? Do they evolve and change from cycle to cycle of a Universe in a history that has neither beginning nor end, ranging over all possibilities, spawning a multiverse of possible worlds, each consistent in its own way, but most devoid of life and unconscious of their own existence?

These are big questions but they grew from little questions. Step by step we have enlarged our view of physical reality, deepened the network of links between parts of it that were superficially different, and found the Universe to be fashioned by nothing more than numbers. And numbers are things we understand, in part. For some, this might be a disappointment. But although the constants of Nature are numbers, they are not just numbers and they are not only numbers. They are the barcodes of ultimate reality, the pin numbers that will unlock the secrets of the Universe — one day.

Notes

'"I do want you to meet Miss Leighton-Buzzard," said Mrs Bovey-Tracey, asking me to dinner the other day. "She's such an interesting woman, and most unusual. She *doesn't write* you know."'

William Plomer,
Electric Delights

chapter one

Before the Beginning

1. H. Mankell, *Sidetracked*, Harvill Press, London, 2000, p. 270.
2. John Donne, 'Sermon, Easter Day 25th March, 1627', in *The Complete Poetry and Selected Prose of John Donne*, ed. C.M. Coffin, Modern Library, New York, 1952, p. 536.
3. B. Appleyard, *Understanding the Present: Science and the Soul of Modern Man*, Doubleday, London, 1992 and V. Havel, Philadelphia Liberty Medal Address, 4 July 1994. www.hrad.cz/president/havel/speeches/1994/0407_uk.html. Havel appears to equate science with technology and thus holds it responsible for all the undesirable things that technology did to people and the environment in the communist states of Eastern Europe.

chapter two

Journey Towards Ultimate Reality

1. Alan Bennett, *Forty Years On*, Faber, London, 1969.
2 Mars Climate Orbiter Mishap Investigation Board Phase I Report, November 10, 1999, available online at ftp: //ftp.hq.nasa.gov/

pub/pao/reports/1999/MCO_report.pdf. The quote is from page 6 of the Executive Summary.

3. Mars Climate Orbiter Mishap Investigation Board Phase I Report, Appendix, p. 37.

4. House Science Committee Chairman F. James Sensenbrenner, Jr, issued a two-word press statement after hearing the news: 'I'm speechless.'

5. An interesting example is provided by the creation of the railway system in Britain. This required that time standards in distant cities had to agree.

6. J. Rivers, *An Audience with Joan Rivers*, London Weekend Television broadcast (1984).

7. A.E. Berriman, *Historical metrology*, Dent, London, 1953.

8. There was a strange attempt to decimalise time as well as units of mass and length. There was an official decree on 24 November 1793 which introduced the new 'revolutionary calendar' which divided months into three 10-day cycles called *décades*. This left the years with 5 special 'spare' days (six in leap years), to be taken after the last month of summer. The system was similar to that used by the ancient Egyptians and it had the ulterior motive of abolishing religious observance of traditional holy days of the week. The innovation failed miserably and the 7-day week was officially reinstated by Napoleon in September 1805. For a more detailed account see J.D. Barrow, *The Artful Universe*, Oxford University Press, London, 1995, p. 159.

9. M. Gläser, *100 Jahre Kilogrammprototyp*, Braunschweig, Physikalisch-Technische Bundesanstalt, 1989.

10. Named from the Greek 'metron' meaning a measure.

11. Originally, Talleyrand had proposed a natural unit of length based on the length of a pendulum that would swing with a period of one second at a latitude of 45 degrees on the Earth's surface.

12. It was of rectangular cross-section 25.3 mm x 4 mm, made of platinum; see T. McGreevy, *The Basis of Measurement*, vol. I, Picton

Publishing, Chippenham, 1995, pp. 148–9.

13. The Royal Society of London did not respond to an invitation to meet with the French Academy of Sciences to agree an international system.

14. It was manufactured by Johnson, Matthey & Co. in London in 1879 together with two copies.

15. Of course, there is the question of how accurately the standard mass is actually known. The prototype's mass is determined to be equal to one kilogram with a measurement uncertainty of 0.135 milligrams. The British standard is to an accuracy of 0.053 milligrams and the American to 0.021 milligrams.

16. M. Kochsiek and M. Gläser, eds, *Comprehensive Mass Metrology*, Wiley-VCH, Berlin, 2000, p. 64.

17. *Things you ought to know!*, Stoney Evans & Co., Rawdon, p. 9 (undated).

18. In 1800 industry would not have required lengths to be determined more accurately than about 0.25 mm, by 1900 the requirement had tightened to about 0.01 mm, by 1950 to 0.25 microns, by 1970 to 12 nanometres. Today, nanotechnologies are manipulating the structure of individual atoms.

19. Note that several nineteenth-century scientists, for example Lord Kelvin, used the term 'metric system' to describe any system of weights and measures at all because the Greek *metron* meant simply measure. They used the term 'decimal system' for what we would now term the metric system based on the metre as a unit of length.

20. J.C. Maxwell, Presidential Address to the British Association for the Advancement of Science, 1870, quoted in B. Petley, *The Fundamental Physical Constants and the Frontier of Measurement*, Adam Hilger, Bristol, 1985, p. 15. It is clear that Maxwell here used 'molecule' where we now use the term 'atom'.

21. The suggestion of using the wavelength of light from specified atomic transitions to define length appears to have first been made in 1827 by a French scientist, J. Babinet, but the equipment needed

to carry this out was not available before his death in 1872.

22. Later the defining wavelength was changed to that of the light emitted when a transition occurs between two energy levels in the krypton-86 atom to allow greater accuracy of measurement.

23. Cadmium can be identified by the number of protons and neutrons that form its nucleus.

24. I.B. Singer, *A Crown of Feathers*, Farrar, Straus & Giroux, New York, 1970, p. 47.

25. G. Johnstone Stoney, *Philosophical Magazine* (series 5), 11, 381 (1881). This article records the material presented at the BAAS Meeting in Belfast in 1874. It is also printed in the *Scientific Proceedings of the Royal Dublin Society*, 3, 51 (1883). This is based on the material presented at a talk on 16 February 1881. The significance of this work was stressed in early editions of the *Encyclopaedia Britannica* by Millikan, who wrote the entry on the 'Electron'.

26. *Notes and Records of the Royal Society*, vol. 29 (October 1974), Plate 14. Reproduced by permission of the Royal Society Library.

27. Stoney wrote in August 1874 in his paper 'On the Physical Units of Nature' that 'Nature presents us, in the phenomenon of electrolysis, with a single definite quantity of electricity which is independent of the particular bodies acted on. To make this clear I shall express "Faraday's Law" in the following terms, which, as I shall show, will give it precision, viz.: – *For each chemical bond which is ruptured within an electrolyte a certain quantity of electricity traverses the electrolyte which is the same in all cases*. This definite quantity of electricity I shall call E_1. If we make this our unit quantity of electricity, we shall probably have made a very important step in our study of molecular phenomena.'

28. D.L. Anderson, *The Discovery of the Electron*, Van Nostrand, Princeton NJ, 1964; I.B. Cohen, 'Conservation and the concept of electric charge: an aspect of philosophy in relation to physics in the nineteenth century', in *Critical Problems in the History of Science*, ed. M. Clagett, University of Wisconsin Press, Madison, 1959.

29. *Scientific Transactions of the Royal Dublin Society* IV series 11 (1891).

30. Stoney had the idiosyncrasy of using the suffix '-ine' in many descriptions of units. For example, he refers to the metre as 'length-ine, or unit of length', gram as 'massine, or unit of mass, second as timine, or unit of time'; see *Scientific Proceedings of the Royal Dublin Society*, 3, 51 (1883).

31. Stoney's name 'electron' was adopted in preference to the label 'corpuscle' which its discoverer, J.J. Thomson, wanted to call it.

32. S. Turing, *Alan Turing*, Heffers, Cambridge, 1959.

33. Other members included Maxwell and William Thomson (later Lord Kelvin); see J.G. O'Hara, 'George Johnstone Stoney, F.R.S., and the concept of the electron', *Notes and Records of the Royal Society of London*, 29, 265 (1974). Stoney's prediction of a basic electric charge does not seem to have attracted the degree of attention it deserved. This is clear from the fact that in October 1894 he can be found writing a letter to the editors of the *Philosophical Magazine*, a leading scientific journal of the day, complaining that Ebert, a recent author in the magazine, had stated that 'Von Helmholtz . . . was the first to show that . . . there must . . . be a minimum quantity of electricity . . . which like an electrical atom is no longer divisible.' Stoney draws attention to his prior lecture and articles of 20 years earlier, see *Philosophical Magazine* series 5, 38, 418 (1894) which can be viewed at http://dbhs.wvusd.k12.ca.us/ Chem-History/Stoney-1894.html.

34. Einstein's general theory of relativity, which extends Newton's theory of gravity to cope with situations where gravity is very strong and motion can take place at the speed of light, preserves the special status of G. The defining constant of the theory is G/c^4 which emphasises its relativistic aspect.

35. There was a period in the early 1960s when the possibility that G was falling with time was taken very seriously by astronomers because the predictions of Einstein's theory of general relativity about the effects of the Sun's gravity seemed to conflict with the

amount observed. The American physicists Carl Brans and Robert Dicke developed a generalisation of Einstein's theory in which G could vary in space or time. This theory is still very important as a tool for making predictions about the consequences of varying G which can be checked against observations. The motivation for Brans and Dicke's development disappeared within a decade. The apparent disagreement between Einstein's theory and the observations was found to be due to inaccuracies in determining the diameter of the Sun due to turbulent activity on its surface. When this was accounted for the theoretical predictions agreed very precisely with the observations.

36. *Scientific Proceedings of the Royal Dublin Society*, 3, 53 (1883).

37. He divides the quantity of electricity required for the electrolysis of 1 cc of hydrogen by the number of hydrogen atoms in 1 cc, which is given by Avogadro's number. Robert Millikan's *Encyclopedia Britannica* (1926–36) article on the Electron, written in 1926, credits Johnstone Stoney's 1881 paper with the first calculation of the expected charge on one electron.

38. Note that this paper was written before modern CGS units for electrical quantities were introduced and Ampères now measure electric current rather than charge (= current × time). Stoney's value for e corresponds to 10^{-11} CGS units.

39. K. Wilber, *Quantum Questions: Mystical Writings of the World's Great Physicists*, New Science Library, Boston, 1985, p. 153.

40. See for example the article 'The Mystery of our Being' and interview with Planck in the collection edited by K. Wilber, *Quantum Questions: Mystical Writings of the World's Great Physicists*, New Science Library, Boston, 1985, chapter 17. They are extracted from his book *Where is Science Going?*, Norton, New York, 1932. See also M. Planck, 'Religion and Natural Science', in *Scientific Autobiography and Other Papers*, Philosophical Library, New York, 1949.

41. Letter to I. Rosenthal-Schneider (30 March 1947), German original and English translation in I. Rosenthal-Schneider, *Reality and*

Scientific Truth, Wayne State Press, Detroit, 1980, pp. 56–7. Rosenthal-Schneider had asked him about the general search for links between constants and Eddington's attempts to do this in particular.

42. M. Planck, 'Über irreversible Strahlungsvorgänge', *S.-B. Preuss Akad. Wiss.* 5, 440–480 (1899). This is also published as *Ann. d. Physik* I, 69 (1900). *Theorie der Warmestrahlung*, Barth, Leipzig, 1906. Engl. translation *The Theory of Heat Radiation*, trans. M. Masius, Dover, New York, 1959.

43. These quantities are defined in an identical discussion which appears in his papers of 1899 and 1900 and were publicised in a series of lectures delivered in Berlin in 1906/7. These lectures were later published as *Theorie der Warmestrahlung*, where the same discussion of natural units ('Natürliche Masseinheiten') appears again.

44. Planck used different symbols for these constants: with f for our G, b for our h, and a for our k. We have used the modern symbols. The symbol G appears to have been introduced for the gravitation constant by A. König and F. Recharz, 'Eine neue Methode zur Bestimmung der Gravitationsconstante', *Ann. Physik. Chem.* 24, 664–8 (1885).

45. Boltzmann's constant is not truly fundamental like G, h, and c. It is just a factor to convert energy units into temperature units.

46. The reason for the co-existence of Stoney and Planck natural units of mass, length and time, each differing in value by a small factor is that the combination e^2/hc is a dimensionless constant of Nature roughly equal to 1/860 using currently determined values of the constants. Therefore, if e^2 is just replaced by hc in Stoney's units of mass, length and time we get Planck's units up to a numerical factor given by the square root of 860. A Stoney natural unit of temperature can be created in the same way.

47. He went further to define units in which 'we now choose the natural units so that in the new system of measurement each of

the four preceding constants assumes the value 1'. This corresponds to measuring all masses, lengths, times, and temperatures in Planck's units.

48. P. Drude, 'Über Fernewirkungen', *Ann. der Physik* 62, i–xlix (supplement) (1898). This was developed further in his textbook on optics published in 1900 and translated by Mann and Millikan as *The Theory of Optics*, Longmans, Green, New York, 1902, see pp. ix, 527.

49. A few years later Drude supports a choice that is really the same as Planck's. He uses c, G and the two radiation constants that define black body radiation. These are reducible to k, Boltzmann's constant and h, Planck's constant; see P. Drude, *The Theory of Optics*, Longmans, Green, New York, 1902, p. 527, where he references Planck's discussion of 1899. He says that 'The absolute system is then obtained from the assumption that the constant of gravitation, the velocity of light, and the two constants . . . in the law of radiation all have the value 1.'

50. According to Focken, writing 1953, Eddington claimed that the Planck length must be the key to some essential structure because it is so much smaller than the radii of the proton and electron. Focken does not reference Eddington's claim but he probably refers to a report on general relativity prepared for the Physical Society of London in 1918; see A.S. Eddington, *Report on the Relativity Theory of Gravitation*, Physical Society of London, Fleetway Press, London, 1918. On the last page of that report, which later evolved into his text on relativity (A.S. Eddington, *The Mathematical Theory of Relativity*, Cambridge University Press, Cambridge, 1923) Eddington derives Planck's natural unit of length and his report contains the remarkable statement that 'There are other natural units of length – the radii of positive and negative unit electric charges – but these are of an altogether higher order of magnitude . . . no theory has attempted to reach such fine-grainedness. But it is evident that this length must be the key to some essential

structure. It may not be an unattainable hope that some day a clearer knowledge of the processes of gravitation may be reached; and the extreme generality and detachment of the relativity theory may be illuminated.' Percy Bridgman also pointed out that the huge value of the Planck temperature, even by astrophysical standards, indicated that it might be associated with some new and fundamental level of cosmic structure; see P.W. Bridgman, *Dimensional Analysis*, Yale University Press, New Haven, 1920.

51. M. Planck, *Scientific Autobiography and Other Papers*, transl. F. Gaynor, Williams & Norgate, London, 1950, p. 170.

52. A. Michelson, public lecture at the University of Chicago, cited in *Physics Today* 21, 9 (1968) and *Light Waves and their Uses*, University of Chicago Press, 1961.

53. O. Wilde, *Phrases and Philosophies for the Use of the Young*, 1894, first published in December 1894 in the Oxford student magazine *The Chameleon*; see *The Portable Oscar Wilde*, eds R. Aldington and S. Weintraub, Viking, New York, 1976.

chapter three

Superhuman Standards

1. A. Conan Doyle, 'The Bruce-Partington Plans', *His Last Bow*, Oxford University Press edition, Oxford, 1993, p. 38; first published as a story in *Strand Magazine* in 1907 and in book form by John Murray, London in 1917.

2. S.W. Hawking and W. Israel, *Einstein: A Centenary Volume*, Cambridge University Press, 1987, p. 128.

3. The 'vacuum' is important. Light moves more slowly in a medium and it is possible for motion to occur in a medium at a speed that exceeds the speed of light in that medium. When it happens a burst of radiation is formed (almost like a sonic boom when going faster than the speed of sound) called Cerenkov radiation, after the Russian physicist who discovered this process. It is very

useful when detecting fast-moving cosmic ray particles coming in from space. Space is for all practical purposes a vacuum and so if you make the incoming particles, which are moving very close to the speed of light in vacuum, enter a medium like water, they will find themselves moving faster than the speed of light in the medium and emit Cerenkov radiation which is easy to detect.

4. See J.D. Barrow, *Theories of Everything*, Oxford University Press, London, 1991.

5. Einstein's quest for a unified field theory amounted only to seeking a way to unite gravity and electromagnetism. He seemed to take no interest in the weak force of radioactivity and the strong nuclear forces. It could be said that his programme for unification was only playing with half the pieces of the puzzle. In 1980 I mentioned this to the mathematical physicist Abraham Taub in Berkeley because Taub had worked closely with John von Neumann at Princeton and had had contact with Einstein there as well. He told me that he once heard this objection put to Einstein who said that he believed that ultimately the weak and strong forces would be shown just to be aspects of the electromagnetic force. This was a prescient comment since we believe that the electromagnetic and weak forces are unified in the well-tested Weinberg-Salam theory, while theories which add the strong force also exist but await decisive observation test.

6. Einstein liked to evaluate entire theories by the 'strength' of their equations (see for example the 5th edition of his book *The Meaning of Relativity*, Methuen, London, 1955). This is just the number of pieces of information that can be freely and independently inputted into the equations, what mathematicians called the 'initial data'. Einstein extended this measure of a theory's tightness to the constants of Nature defining its solutions as well.

7. I. Rosenthal-Schneider, *Reality and Scientific Truth: Discussions with Einstein, von Laue, and Planck*, Wayne State University Press, Detroit, 1980, p. 32.

8. Ibid., frontispiece.

9. Ibid., p. 34.

10. Ibid., p. 38.

11. For instance, if we calculate the circumference of a circle of radius R we find it to be equal to $2\pi R$. The factor of 2π is one of these ubiquitous 'basic' numbers.

12. It means that they can use dimensional analysis of physical problems to guess at the form of exact equations.

13. If you choose e, h, and c then there is e^2/hc. This fact was exploited by Hartree who created a set of units for atomic physics investigations which employ e, h, c and the mass of the electron m_e.

14. It is just equal to the mass ratio $(m_{pr}/m_{pl})^2 \approx (10^{-24}\,\text{gm}/10^{-5}\,\text{gm})^2 \approx 10^{-38}$ where m_{pl} is Planck's fundamental mass.

15. Einstein points out that this is not a procedure that can be carried out in full for all of physics at present because we do not know all the laws and formulae that govern it.

16. At the time when Einstein was corresponding on these matters the only ideas that existed about why the constants took the values that they did were those of Eddington, which were not greeted with much enthusiasm by other physicists. Einstein commented in a later letter of 23 April 1949 to Rosenthal-Schneider on Eddington's numerology. She had written to him asking if she could quote from his letters in the Library of Living Philosophers volume about Einstein to which she was invited to contribute an article. He replies: 'You may make use of my remarks in your treatise; it must however be said that these are by no means categorical assertions, but merely conjectures grounded on intuition. Eddington made many ingenious suggestions, but I have not followed them all up. I find that he was as a rule curiously uncritical toward his own ideas. He had little feeling for the need for a theoretical construction to be logically very simple if it is to have any prospect of being true.'

17. G. Gamow, 'Any physics tomorrow?', *Physics Today*, January 1949.

18. J.R.R. Tolkien, *The Lord of the Rings*, part I, *The Fellowship of the Ring*, Unwin, London, 1954.

19. When the universe has age T, the visible universe has size cT where c is the speed of light.

20. Roughly a mass of 10^{-24} gm in every volume of $(10^{-8}$ cm$)^3$. This is roughly the density of water, 1 gm per cc, and most other solids, liquids and gases don't vary very much from this in density.

21. Based on diagram from B.J. Carr and M.J. Rees, 'The anthropic principle and the structure of the physical world', *Nature* 278, 605 (1979).

22. British TV programme *They Think It's All Over*, 5 December 1999.

23. Quoted in T.A. Bass, *The Predictors*, Penguin, London, 2000, p. 172.

24. This is an idealisation. The most distant stars are not perfectly at rest with respect to us in this sense but their motions are imperceptibly small. One of the achievements of Einstein's theory of gravity and motion, which superseded Newton's, was to do away with this imaginary background of 'absolute space'. Newton himself was criticised by philosophers like Bishop Berkeley for introducing such a concept. Newton was aware of its weaknesses but recognised its usefulness in expressing a theory of motion that was extremely accurate in describing local motions.

25. Rotational motion is always accelerated motion. Even if the speed of motion is constant the direction of motion must continually change to maintain the circular motion. Hence the velocity is always changing. This is what is meant by acceleration.

chapter four

Further, Deeper, Fewer: The Quest for a Theory of Everything

1. R.P. Crease, 'Do physics and politics mix?', *Physics World*, Feb. 2001, p. 17.

2. Quoted in C. Pickover, *The Loom of God*, Plenum, New York, 1997, p. 26.

3. J.D. Barrow, *The Universe that Discovered Itself*, Oxford University Press, London, 1990, discusses the development of the concept of 'laws' of Nature in greater detail.

4. We do not expect every possible outcome of the laws of Nature to exist in reality. Thus the real world is a subset of all possible worlds. It is an interesting question as to what is the objection to a world in which there are logical inconsistencies in the outcomes of the laws but they are not manifested in any actual outcomes.

5. Quoted in J.A. Paulos, *I Think, Therefore I Laugh*, Columbia University Press, New York, 1985, p. 35.

6. See, for example, the modern reprint G. Gamow, *Mr. Tompkins in Paperback*, Cambridge University Press, Cambridge, 1949. An updated and extended version of some of Mr Tompkins's educational experiences can be found under the editorship of Russell Stannard.

7. The speed of light was first brilliantly deduced by the Danish astronomer Olaf Roemer in 1676. He noticed that the intervals of time that passed between the eclipses of one of Jupiter's moons lengthened when the Earth was moving away from Jupiter but shortened when it approached Jupiter. He found an average time difference of 996 seconds between eclipses from many observations made over the course of a year. Roemer then attributed this time difference to the fact that light had a finite speed. Thus, he argued, it must require 996 seconds for light to cross a distance equal to the diameter of the Earth's orbit. This distance was accurately known, even then, and allowed him to obtain a very good estimate of the speed of light.

8. Gamow, *Mr. Tompkins in Paperback*, p. I.

9. Gamow has to be allowed some artistic licence here. As we explained in the last chapter, just varying the dimensional

constants of Nature, like the speed of light, does not lead to any observational differences in the behaviour of the world if other constants also vary so that all the dimensionless constants stay the same.

10. The non-zero value of h is important for the stability of matter. If the energy of an atom could change by an arbitrarily small amount then all atoms would soon become very different. The buffetings of other atoms and radiation would change their energy levels all the time. The constant h is large enough that atoms need quite a big 'kick' before they can be knocked into the next allowed level.

11. According to L.B. Okun, this representation of the constants was introduced first by the Russian physicist Matveí Bronstein in the early 1930s. Unfortunately, Bronstein was murdered by Stalin in 1938 when only 32 years old. A biography exists (in Russian) by G.E. Gorelik and V. Ya. Frenkel, *Matveí Petrovich Bronstein*, Nauka, Moscow, 1990.

12. David Singmaster reported in M. Stueben and D. Sandford, *Twenty Years before the Blackboard*, Math. Assoc. of America, Washington DC, 1998, p. 95.

13. It is not possible so far to predict what should remain after the final explosion. Many different suggestions have been made, ranging from nothing at all, to a hole in space and time, a wormhole into a new universe, or just a finite stable mass.

14. We don't know, for example, if the fine structure constant is a rational or an irrational number.

15. D.M. Wilson, *Awful Ends: The British Museum Book of Epitaphs*, British Museum Publications, London, 1992, p. 87.

16. C. Butler, *Number Symbolism*, Routledge & Kegan Paul, London, 1970.

17. The nth triangular number equals $n(n + 1)/2$.

18. In general, the n^2 equals the sum of the first n odd numbers, starting with 1.

19. Alexander of Aphrodisias (a commentator on Aristotle) in his *Metaphysica*, 38, 10 quoted by W. Guthrie, *History of Greek Philosophy*, vol. I, Cambridge University Press, Cambridge, 1962, pp. 303–4.

20. All perfect numbers can be expressed as $2^N(2^{N+1} - I)$ for special values of N. The great Swiss mathematician Leonhard Euler showed that all even perfect numbers have this form if $2^N - I$ is a prime number. No one knows if odd perfect numbers exist.

21. Prime numbers, like 7 and 23, have no divisors other than themselves and I. Euclid showed that there are infinitely many of them by a most beautiful argument. Suppose there are only a finite number of them. Multiply them all together and add one. Then this number is not divisible by any of your assumed finite list of primes because there is always a remainder of I. Therefore either this number is a prime or it is divisible by a prime that is bigger than the last one in your original list. Either way, this contradicts the original assumption that the list of primes was finite. Hence the number of prime numbers cannot be finite.

22. More than a thousand amicable numbers have been discovered. The next largest are 1184 and 1210, 2620 and 2924, 5020 and 5564, 6232 and 6368, 10744 and 10856.

23. In the book of Genesis 32, verse 14, the amicable number 220 appears when Jacob gives a gift of 220 goats to Esau. The implication is of a relationship that would be sealed by the reciprocal gift of 284 things.

24. From Trachtenberg, *Jewish Magic and Superstition*, quoted by C. Pickover, *The Loom of God*, Plenum, New York, 1997, p. 80.

25. Theon of Smyrna, On the Tetraktys and the Dead, quoted by C. Butler, *Number Symbolism*, Routledge & Kegan Paul, London, 1970, p. 9.

26. H. Weber, *Lehrbuch der Algebra*, vol. 3, Chelsea, New York, 1908, section 125. This example is quoted by I.J. Good in an unpublished Dept. Statistics, Virginia Polytechnic Inst. Technical Report, *Physical Numerology*, 30 Dec. 1988, p. I.

27. This was part of an April Fool's Day hoax in the Martin Gardner column in the April 1975 edition of *Scientific American*, p. 127. (The hoax was revealed in the July 1975 issue, p. 112). One can prove that there exist rational numbers equal to an irrational number raised to an irrational power but as far as I know no explicit examples are known. The proof is a beautiful example of a non-constructive proof. Consider the number x equal to $\sqrt{2}$ raised to the power of $\sqrt{2}$. This number is either rational or irrational. If it is rational we have proved what we are seeking, so assume it is irrational. Raise it to the power of $\sqrt{2}$ again and we have that $x^{\sqrt{2}} = (\sqrt{2})^{\sqrt{2} \times \sqrt{2}} = (\sqrt{2})^2 = 2$ which is rational and equal to an irrational raised to an irrational power, by assumption!

28. The Authorised Version was the outcome of the Hampton Court Conference of 1604 which was convened by James I to bring together various High and Low Church parties. The resulting 'Authorised Version' (although it was not actually 'authorised' in any official way at all) appeared in 1611. It is based largely on the translated texts of William Tyndale with material from John Wyclif. William Shakespeare lived from 1564 to 1616.

29. The first and last stanzas of Psalm 46 read (with the 46th words from the start and finish capitalised!):

God is our refuge and strength,
A very present help in trouble.
Therefore will not we fear, though the earth be removed,
And though the mountains be carried into the midst of the sea;
Though the waters thereof roar and be troubled,
Though the mountains SHAKE with the swelling thereof
. . .
He breaketh the bow, and cutteth the SPEAR in sunder;
He burneth the chariot in the fire.
'Be still, and know that I am God:
I will be exalted among the heathen, I will be exalted in the earth.'

The Lord of hosts is with us;
The God of Jacob is our refuge.

30. G.N. Lewis and E.Q. Adams, *Phys. Rev.* 3, 92 (1914).

31. A.S. Eddington, *Proc. Roy. Soc.*, A 122, 358 (1930). Notice that Eddington at this time believed that $1/\alpha$ was a whole number. At the time this was a possibility given the experimental uncertainties in its measurement.

32. A.M. Wyler, C. Rendus, *Acad. Sci.*, Paris B 269, 743 (1969) and B 271, 186 (1971).

33. H. Aspden and D.M. Eagles, *Phys. Lett.*, A 41, 423 (1972).

34. C. Pickover, *Computers and the Imagination*, St Martin's Press, New York, 1991, p. 270.

35. B. Robertson, *Phys. Rev. Lett.*, 27, 1545 (1971).

36. T.J. Burger, *Nature* 271, 402 (1978).

37. W. Heisenberg, letter to Paul Dirac, 27 March 1935, quoted in H. Kragh, *Dirac: A Scientific Biography*, Cambridge University Press, Cambridge, 1990, p. 209.

chapter five

Eddington's Unfinished Symphony

1. A.S. Eddington, *The Expanding Universe*, Cambridge University Press, Cambridge, 1933, p. 126.

2. R. Scruton, *The Intelligent Person's Guide to Philosophy*, quoted in the *Times Higher Educational Supplement*, 4 May 2001, p. 19.

3. A.V. Douglas, *The Life of Arthur Stanley Eddington*, Nelson, London, 1956, Plate II.

4. A.V. Douglas, *The Life of Arthur Stanley Eddington*, Nelson, London, 1956; H.C. Plummer, Arthur Stanley Eddington 1882–1944, Obituary Notices of Fellows of the Royal Society, V, 1945–8, pp. 113–125; C.W. Kilmister, *Men of Physics: Sir Arthur Eddington*, Pergamon, Oxford, 1966; E.T. Whittaker, Arthur Stanley

Eddington, *Dictionary of National Biography*, 1941–50, pp. 230–3; W.H. McCrea, 'Recollections of Sir Arthur Eddington', *Contemporary Physics* 23, 531–40 (1982).

5. D.L. Sayers, *Have His Carcase*, Victor Gollancz, London, 1932. This title is cockney rhyming slang for Habeus Corpus, the Act of Parliament that requires that the accused be presented with the evidence against him or her before a judge. The quotation is from p. 206 of the 1948 impression.

6. His successor, R.O. Redman, wrote that, 'Eddington enjoyed crowds. At one time, on every Saturday in the football season, he would be off by himself, not to the Rugger usually favoured by Cambridge dons, but to a professional Soccer game, with its large crowd of working-class fans', quoted in A.V. Douglas, *Arthur Stanley Eddington*, Nelson, London, 1956, p. 122.

7. A. S. Eddington, *The Philosophy of Physical Science*, Cambridge University Press, Cambridge, 1939, p. 58.

8. Following Eddington's early death in November 1944, the manuscript was published posthumously as *Fundamental Theory* by Cambridge University Press in 1946 under the editorship of Eddington's friend and former mentor E.T. Whittaker. The title was chosen by Whittaker. Subsequently, attempts were made to try to make plain the methodology of Eddington's work by N.B. Slater in *Development and Meaning of Eddington's Fundamental Theory*, Cambridge University Press, Cambridge, 1957 which was reviewed at length by A. Taub, *Mathematical Reviews* 11, 144 (1950). C. Kilmister and B.O.J. Tupper, *Eddington's Statistical Theory*, Clarendon Press, Oxford, 1962.

9. A.S. Eddington, Address to the British Association, 1920, *Observatory* 43, 357–8 (1920).

10. For he said, 'An electron would not know how large it ought to be unless there existed independent lengths in space for it to measure itself against.' A.S. Eddington, *The Mathematical Theory of Relativity*, Cambridge University Press, Cambridge, 1923, p. 33.

11. Actually only the number in the part of the Universe which is visible in principle, given the finiteness of the speed of light. The number of protons in the entire Universe could be infinite or finite depending upon the global geometry of space.

12. He had estimates of the density and size of the Universe from astronomy and so could calculate the mass by multiplying them together. By dividing this total mass by the mass of a proton he gets the number of protons in the Universe. This would have taken him about thirty seconds to calculate. What took the boat trip was the tedious task of expressing the answer as a single whole number.

13. A.S. Eddington, *New Pathways in Science*, Cambridge University Press, Cambridge, 1935, p. 232.

14. Eddington, ibid., p. 233 and p. 234.

15. Eddington, ibid., p. 234.

16. Although Eddington was greatly preoccupied with these 'large' numbers of the order of 10^{40} and powers thereof, he was not the first person to notice their appearance in combinations of the constants of Nature. That discovery was first made by Hermann Weyl in 1919. He noted that, 'It is a fact that pure numbers appear with the electron, the magnitudes of which are totally different from 1; so for example, the ratio of the electron radius to the gravitational radius of its mass is of order 10^{40}; the ratio of the electron radius to the world radius may be of similar proportions', *Ann. Physik* 59, 129 (1919) and *Naturwissenschaften* 22, 145 (1934).

17. A.S. Eddington, *Philosophy of Physical Science*, Cambridge University Press, Cambridge, 1939, p. 69.

18. They are $[136 \pm \sqrt{18456}] \div 20 = [136 \pm 135.85286] \div 20 = 13.5926$ or 0.007357 so the ratio is 1847.57.

19. Eddington, *New Pathways in Science*, Cambridge University Press, Cambridge, 1935, p. 251. His explanation, was as follows: 'By a rather precarious argument it seems likely that when a number of electric charges form a perfectly rigid system, $1/137$ of their mass

is lost. Since the atomic nucleus is approximately rigid, this should give an approximate determination of the 'packing fraction', *Proc. Roy. Soc.* A 126, 696 (1930).

20. V.A. Fock, quoted by George Gamow in *Biography of Physics*, Harper & Row, New York, 1961, p. 327. Fock was an influential Soviet physicist who attempted to make Einstein and his work politically acceptable during the Stalinist period. In particular, he renamed Einstein's relativity theory 'the theory of invariance' in order to counter the accusation that the theory was in some way opposed to the absolute truth of dialectical materialism. His notable text on Einstein's theory of general relativity, *The Theory of Space, Time and Gravitation* published by Pergamon (Oxford, 1959) contains a famous foreword that the book was only possible because of the positive influence of dialectical materialism.

21. G. Beck, H. Bethe and W. Riezler, *Naturwissenschaften*, 19, 29 (1931). The translation here is one by Max Delbrück, in *Cosmology, Fusion, and Other Matters*, ed. F. Reines, Adam Hilger, Bristol, 1972. It is worth noting that at this time there was some serious consideration of the possibility that the fine structure constant might be linked to the temperature concept. Paul Dirac was interested in this possibility and it was considered by Heisenberg who mentions his disenchantment with it in a letter to Dirac a few years later, writing on 27 March 1935 that 'I don't believe at all any more in your conjecture that the Sommerfeld fine structure constant may have something to do with the concept of temperature ... Rather, I am firmly convinced that one must determine e^2/hc within the whole theory', in H. Kragh, *Dirac: A Scientific Biography*, Cambridge University Press, Cambridge, 1990, p. 209.

22. Arnold Sommerfeld had introduced α into physics, calling the symbol a new 'Abkürzung' (abbreviation), in *Sitz. Ber. Bayer. Akad. Wiss.*, p. 459 (1915).

23. According to Delbrück, ref. 21, this was A.V. Das.

24. Born referred to this short book as his 'anti-Eddington and Milne

essay' in a letter to Einstein the following year, see M. Born, *Albert Einstein – Max Born, Briefwechsel 1916–1955*, Rowohlt, Hamburg, 1972, letter of 10 October 1944.

25. M. Dresden, *H.A. Kramers: Between Tradition and Revolution*, Springer, New York, 1987, p. 518.

26. J.D. Barrow and F.J. Tipler, *The Anthropic Cosmological Principle*, Oxford University Press, London, 1986, p. 231.

27. U. Dudley, *Numerology, or what Pythagoras wrought*, Math. Assoc. of America, Washington DC, 1997, p. 7.

28. J. Jeans, *The Growth of Physical Science*, Cambridge University Press, Cambridge, 1947, p. 357.

29. Letter to Dingle quoted in J.G. Crowther, *British Scientists of the Twentieth Century*, Routledge & Kegan Paul, London, 1952, p. 194.

chapter six

The Mystery of the Very Large Numbers

1. P. Valéry, *Variété* IV.

2. This is defined as the region from which light has had time to travel since the expansion apparently began. It is a sphere of radius approximately 13 billion light years centred upon ourselves.

3. Because each force falls off inversely as the square of the distance of their separation.

4. This is roughly equal to its energy multiplied by its age.

5. In 1980 there was considerable interest in the possibility that the proton might be unstable with a half-life close to about 10^{31} years (for a while there were claims, ultimately unsubstantiated, that this decay had been detected). I pointed out at the time that the ratio of this predicted lifetime to the fundamental Planck time was about 10^{80}, see J.D. Barrow, 'The Proton Half-life and the Dirac Hypothesis', *Nature*, 282, 698–9, (1979).

6. Remark made by Gamow to Niels Bohr on seeing Dirac's paper on the Large Numbers Hypothesis in *Nature*, G. Gamow, 'History

of the Universe', *Science* 158, 766–9 (1967). Dirac had got married just a month before the paper was written.

7. P.A.M. Dirac, 'A New Basis for Cosmology', *Proc. Roy. Soc.* A 165, 199–208 (1938).

8. Dirac remarked that 'Eddington's arguments are not always rigorous . . . [but] 10^{39} and 10^{78} are so enormous as to make one think that some entirely different type of explanation is needed for them.'

9. P.A.M. Dirac, *Nature* 139, 323 (1937) and *Proc. Roy. Soc.* A 165, 199 (1938). Dirac means that any two separate collections of dimensionless constants of Nature must be proportional, where the constant of proportionality must be fairly close to one, say about a tenth or ten, perhaps composed of purely numerical factors like 2 and π. Numerical factors that were very large or small, a million for example, would not be allowed.

10. The conclusion $N \propto t^2$ subsequently led Dirac to conclude (P.A.M. Dirac, *Proc. Roy. Soc.* A 333, 403 (1973)), quite wrongly, that this result required the continuous creation of protons. In fact, all it is telling us is that as the universe ages we are able to see more and more protons coming within our horizon.

11. Of course, this hypothesis is able to tell us why the different collections of constants N_1, N_2 and $\sqrt{N}$ are of similar magnitude but not why the magnitude is now close to 10^{40}.

12. The most vociferous critic was Dingle, who linked together the theories of Milne and Dirac as examples of a combination of 'paralysis of the reason with intoxication of the fancy . . . Instead of the induction of principles from phenomena we are given a pseudoscience of invertebrate cosmythology, and invited to commit suicide to avoid the need of dying.' H. Dingle, 'Modern Aristotelianism', *Nature* 139, 784 (1937).

13. P.A.M. Dirac, 'The Relation between Mathematics and Physics', *Proc. Royal Society* (Edinburgh) 59, 129 (1937).

14. The luminosity of the Sun is proportional to G^7 and the radius of

the Earth's orbit around the Sun is proportional to G^{-1} so the average temperature at the Earth's surface is proportional to $G^{9/4} \propto t^{-9/4}$.

15. E. Teller, *Phys. Rev.* 73, 801 (1948).

16. A change in the value of e does not affect the Earth's orbit around the Sun, while the luminosity of the Sun is proportional to e^{-6} so the average surface temperature of the Earth is proportional to $t^{-3/4}$ and the era of boiling oceans would be shifted too far into the past to be a problem for our biological history.

17. P.A.M. Dirac, letter to Gamow, cited by H. Kragh, *Dirac: A Scientific Biography*, Cambridge University Press, Cambridge, 1990, p. 236, original in Library of Congress, manuscript collection.

18. Six billion years was his estimate of the age of the Universe at that time. We now know this was a significant underestimate because of a miscalibration of the distances to the galaxies which was corrected in 1953.

19. A. Hodges, *Alan Turing: The Enigma of Intelligence*, Hutchinson, London, 1983.

20. J.B.S. Haldane, 'Radioactivity and the Origin of Life in Milne's Cosmology', *Nature* 158, 555 (1944); see also *Nature* 139, 1002 and Haldane's article in *New Biology*, No. 16, eds M.L. Johnson, M. Abercrombie and G.E. Fogg, Penguin, London, 1955.

21. See C. Will, *Theory and Experiment in Gravitational Physics*, Cambridge University Press, Cambridge, 1981, p. 181.

22. R.H. Dicke, 'Principle of Equivalence and Weak Interactions', *Rev. Mod. Phys.* 29, 355 (1957) and *Nature* 192, 440, (1961).

23. E.A. Milne, *Modern Cosmology and the Christian Idea of God*, Oxford University Press, London, 1952, p. 158.

24. By this we mean chemical elements heavier than helium.

25. P.A.M. Dirac, Letter to Heisenberg, 6 March 1967, quoted by L.M. Brown and H. Rechenberg, in B. Kursunoglu and E. Wigner (eds), *Paul Adrien Maurice Dirac. Reminiscences about a Great Physicist*, Cambridge University Press, Cambridge, 1987, p. 148.

26. Dirac to Gamow, 20 November 1967, quoted in H. Kragh, *Dirac:*

A Scientific Biography, Cambridge University Press, Cambridge, p. 238.

27. E. Mascall, *Christian Theology and Natural Science*, Longmans, London, 1956, p. 43. Mascall refers to an 'unpublished paper' by Whitrow. When I asked Professor Whitrow about this in 1979 he wrote back apologetically that 'I have no recollection of what, if anything, happened to the "unpublished paper".'

28. W.C. Fields, *You're Telling Me*, 1934.

29. K. Jaspers, *The Origin and Goal of History*, transl. M. Bullock, Greenwood Press, Westpoint, 1976, p. 237, first published in 1949 as *Vom Ursprung und Ziel der Geschichte*. I am grateful to Yuri Balashov for drawing my attention to this work.

30. It is an interesting coincidence (also partially explained by the fact that we most likely live at the time when the stars shine) that the number of stars in a galaxy is roughly equal to the number of galaxies in the visible Universe. Both number about one hundred billion. In the far future (if there are stars and galaxies) the observable Universe will be bigger and contain more galaxies.

31. This is reported in Albrecht von Haller, *Elementa Physiologiae*, vol. 5, London, 1786, p. 547.

32. These estimates are due to Mike Holderness, 'Think of a Number', *New Scientist*, 16 June 2001, p. 45.

chapter seven

Biology and the Stars

1. D. Adams, *The Restaurant at the End of the Universe*, Pan, London, 1980, p. 84.

2. For pictures of these dramatic events, see http://nssdc.gsfc.nasa.gov/planetary/s19/comet_images.html.

3. Modified version of Figure 8.1 in P.D. Ward and D. Brownlee, *Rare Earth*, Copernicus, New York, 2000, p.165.

4. Ibid., p. 173

5. B. Carter, *Phil. Trans. Roy. Soc.* A 310, 347 (1983).

6. J.D. Barrow and F.J. Tipler, *The Anthropic Cosmological Principle*, Oxford University Press, London, 1986.

7. There is now a vast literature on this so called 'Doomsday argument', see for example J. Leslie, *The End of the World: The Science and Ethics of Human Extinction*, Routledge, London, 1996; H.B. Nielsen, 'On Future Population', *Acta, Phys. Polonica* B 20, 427 (1989); J.R. Gott, 'Implications of the Copernican Principle for our Future Prospects', *Nature* 363, 315–19 (1993) and 'How the Copernican Principle is Consistent with a Bayesian Approach', *Nature* 368, 108 (1994). For a selection of other articles see the website of Nick Bostrum at http://www.anthropic-principle.com/preprints.html.

8. Because there are so many more ways for the two times to be very different than for them to be similar.

9. M. Livio, 'How Rare Are Extraterrestrial Civilizations and When Did They Emerge?' *Astrophys. J.* 511, 429 (1999).

10. J. Laskar and P. Robutel, 'The Chaotic Obliquity of the Planets', *Nature* 361, 608–12; see also J.D. Barrow, *The Artful Universe*, Oxford University Press, London, 1995, pp. 145–9.

11. M. O'Donoghue, quoted in *Playboy* Magazine, 1983.

12. F. Hoyle, *The Black Cloud*, Heinemann, London, 1957.

13. Quoted in the *Observer*, 20 January 2002, p. 26.

14. J.R. Gott, *Time Travel in Einstein's Universe*, Houghton Mifflin, New York, 2001.

15. Ibid., p. 221; first published in *Wall Street Journal*. Reproduced by permission of J.R. Gott.

16. A. Conan Doyle, 'The Final Problem', *The Memoirs of Sherlock Holmes*, Oxford University Press, New York, 1993. 'The Final Problem' was first published in the *Strand* magazine in December 1883 in London and New York.

17. A.N. Whitehead, *Adventures of Ideas*, Cambridge University Press, Cambridge, 1933, part 4, chapter 16.

18. A.R. Wallace, *Man's Place in the Universe*, Chapman & Hall, London,

1903. Page refs are to the 4th edn of 1912.

19. Surprisingly, there had been almost no attempt to create a Newtonian description of the Universe. The notable exception is the remarkable paper by Lord Kelvin (William Thomson) 'On the Clustering of Gravitational Matter in Any Part of the Universe', *Nature* 64, 626 (1901) and *Philosophical Magazine* 3, 1 (1902). This paper is reproduced in its entirety in the article by E.R. Harrison, 'Newton and the Infinite Universe', *Physics Today* 39, 24 (1986).

20. He argued that if there were 10 billion stars the velocities would become too large. In gravitating systems containing a total mass M, radius R and average speed of movement v, these three quantities are generally linked by a relation $v^2 \approx 2GM/R$, where G is Newton's constant.

21. A.R. Wallace, *Man's Place in the Universe*, Chapman & Hall, London, 4th edn, 1912, p. 248.

22. Wallace, ibid., p. 255 and p. 261.

23. Wallace, ibid., p. 256.

24. Wallace, ibid.

25. Wallace, ibid., pp. 256–7.

26. He was particularly struck by the fact that the determination of the speed of light by observations of eclipses of Jupiter's moons matched the value determined terrestrially, concluding that 'These various discoveries give us the certain conviction that the whole material universe is essentially one, both as regards the action of physical and chemical laws, and also in its mechanical relations of form and structure,' Wallace, ibid., p. 154.

27. Wallace, ibid., pp. 154–5.

chapter eight

The Anthropic Principle

1. W.V. Quine, interview for the *Harvard Magazine*, quoted in R. Hersh, *What is Mathematics Really?*, Vintage, New York, 1998, p. 170.

2. H. Pagels, 'A Cozy Cosmology', *The Sciences*, March/April, 34 (1985); G. Kane, M. Perry and A. Zytkow, 'The Beginning and the End of the Anthropic Principle', *New Astronomy* VII, 45–53 (2002).

3. D.A. Redelmeier and R.J. Tibshirani, *Nature* 401, 35 (1999) and *Chance* 13, 8–14 (2000).

4. N. Bostrom, 'Observational Selection Effects and Probability', PhD thesis, see www.anthropic-principle.com/phd/.

5. Another effect is that the next lane appears to be moving faster on a congested motorway even when the average speed of the cars in each lane is the same. This is because the faster-moving cars become more widely spaced while the slower-lane traffic becomes more closely packed.

6. Traffic on Hollywood freeway, © Bettmann/Corbis.

7. E.R. Harrison, *Darkness at Night*, Harvard University Press, Cambridge, MA, 1987, p. 87.

8. G. Santayana, *The Sense of Beauty*, Dover, New York, 1955, first publ. 1896, pp. 102–3.

9. F. Ramsey, *The Foundations of Mathematics and Other Logical Essays*, London, Kegan Paul, Trench and Trubner, 1931, p. 291.

10. This title was invented later, somewhat pejoratively, by Fred Hoyle in a 1949 radio broadcast to emphasise the dramatic beginning required in the usual expanding Universe theory, and published in 1950.

11. The expansion rate of the Universe has units of an inverse time. The inverse of the expansion therefore gives a time which is roughly equal to the age of the Universe in a Big Bang model. In the steady state Universe the inverse of the expansion rate has units of a time but it does not correspond to the true age of the steady state Universe, which is infinite.

12. In fact, Holloway and Moore had reported evidence for an excited state of carbon near 7 MeV in 1940 (*Phys. Rev.* 58, 847 (1940)), and it appears in the nuclear data tables published in *Rev. Mod. Phys.*

20, 23 by a team of which Fowler was a member, but it was not confirmed by subsequent studies by Malm and Buechner, *Phys. Rev.* 81, 519 (1951) and seems to have then been removed from future data tables. I am grateful to Virginia Trimble for this information.

13. F. Hoyle, D.N.F. Dunbar, W.A. Wensel and W. Whaling, *Phys. Rev.* 92, 649 (1953). C.W. Cook, W.A. Fowler and T. Lauritsen, *Phys. Rev.* 107, 508 (1957).

14. This was noticed by E. Salpeter, *Astrophysical Journal* 115, 326 (1952), and G.K. Öpik, *Proc. Roy. Irish Acad.* A54, 49 (1951).

15. F. Hoyle, *Astronomy and Cosmology: A Modern Course*, W.H. Freeman, San Francisco, 1975, p. 402.

16. H. Oberhummer, A. Csótó, and H. Schlattl, *Science* 289, 88 (2000).

17. F. Hoyle, *Galaxies, Nuclei and Quasars*, Heinemann, London, 1965, pp. 159–160.

18. F. Hoyle, ibid., p. 160.

19. F. Hoyle, *Religion and the Scientists*, SCM, London, 1959.

20. Badly engineered features were conveniently overlooked. For an interesting discussion of these see George Williams, *Plan and Purpose in Nature*, Weidenfeld & Nicolson, London, 1996.

21. Charles Darwin was much influenced by the collection of biological Design Arguments, marshalled by authors like William Paley, because he said that they served to line up all the evidence which called for an alternative explanation; see J.D. Barrow and F.J. Tipler, *The Anthropic Cosmological Principle*, Oxford University Press, Oxford, 1986 for a fuller discussion of these developments.

22. All these different influences are discussed systematically in my earlier book *Theories of Everything*, Oxford University Press, Oxford, 1990 and Vintage, New York, 1992.

23. It is important to recognise that this version of the Argument from Design played an important role in lining up countless examples of apparent design in the natural world. It was this that motivated Wallace and Darwin to look for another explanation. Without the parading of the evidence for apparent design there

would have been no focus of attention on it as a problem in need of an explanatory mechanism, see J.D. Barrow and F.J. Tipler, *The Anthropic Cosmological Principle*, Oxford University Press, Oxford, 1986, chap. 2.

24. F. Dyson, *Disturbing the Universe*, Harper & Rowe, New York, 1979.

25. H. Bondi, *Cosmology*, Cambridge University Press, Cambridge, 1952, chapter 13 is devoted to Large Numbers and varying constants.

26. This does not follow. We know that the outcomes of the laws of Nature do not have to possess the same symmetries as the laws themselves. Outcomes are far more complicated, and far less symmetrical than laws.

27. Bondi, ibid., p. 160.

28. B. Carter, 'Large Number Coincidences and the Anthropic Principle in Cosmology', in M.S. Longair (ed.), *Confrontation of Cosmological Theories with Observational Data*, Reidel, Dordrecht, 1974, pp. 291–8.

29. Carter, ibid., p. 292.

30. Whitrow used this argument first to understand why we find space to have three dimensions, as we shall see in a later chapter.

31. Carter was a student and then lecturer in the Department of Applied Mathematics and Theoretical Physics at Cambridge at the time when Dirac was the Lucasian Professor.

32. B. Carter, 'The Anthropic Principle: Self-selection as an Adjunct to Natural Selection', in S.K. Biswaset et al (ed.), *Cosmic Perspectives*, Cambridge University Press, Cambridge, 1988, pp. 187–8.

33. T.S. Eliot, 'The Love Song of J. Alfred Prufrock', *Selected Poems*, Faber and Faber, London, 1994.

34. It only just fails to be bound by about 70 KeV in practice. The significance of this was first pointed out by Freeman Dyson.

35. Adapted from M. Tegmark, *Annals of Physics* 270, 1–51 (1998), using constraints from Barrow and Tipler, op. cit.

36. Adapted from M. Tegmark, *Annals of Physics* 270, 1–51 (1998), using constraints from Barrow and Tipler, op. cit.

37. Woody Allen, quoted in the *Observer* newspaper, 27 May 2001, p. 30.

38. Some biologists would actually define life as anything that evolves by natural selection.

39. This will be the case if the acceleration is caused by the presence of the so-called 'cosmological constant', which represents the vacuum energy of the Universe. It is possible that other forms of matter can mimic a cosmological constant's presence for a finite period of cosmic history before decaying into ordinary forms of matter which do not produce accelerated expansion (see J.D. Barrow, R. Bean, and J. Magueijo, *Mon. Not. R. Astron. Soc.* 316, L41–4 (2000)). If this happens early enough then information processing need not eventually die out.

40. One loophole that might be exploitable in the right type of universe is the possibility that the acceleration is produced by the presence of a new form of matter which might be used as a new form of energy source. This would probably only result in the production of usable energy plus another accelerating constant energy source which could not be mined for energy. Eventually that new source would come to drive the expansion and unstoppable information degradation would commence again.

41. Barrow and Tipler, op. cit., p. 668. Some further discussion has been provided by L. Krauss and G.D. Starkman, *Astrophys. J.* 531, 22–30 (2000).

42. This acceleration may be contributed by a positive cosmological constant, proposed first by Einstein in his original announcement of the general theory of relativity. It is like an additional part to the law of gravity. Unlike the familiar inverse-square law of Newton, this contribution increases linearly with distance. It has a natural interpretation as the vacuum energy of the Universe but its value is very mysterious – 10^{120} times bigger than its value in 'natural' Planck units.

43. K. Gödel, 'An example of a new type of cosmological solution

of Einstein's Field Equations of Gravitation', *Review of Modern Physics* 21, 447–50 (1949).

44. M.R. Reinganum, 'Is Time Travel Impossible? A Financial Proof', *Journal of Portfolio Management* 13, 10–12 (1986).

chapter nine

Altering Constants and Rewriting History

1. D. Adams, *Mostly Harmless*, Heinemann, London, 1992, p. 25.

2. R.A. Heinlein, *The Number of the Beast*, New English Library, London, 1980, p. 14.

3. This is a somewhat hypothetical situation. We might hope to understand why our final theory cannot be changed in any small way without destroying its logical self-consistency, but it is hard to imagine how we could ever know that there was not a completely different self-consistent theory that was not in any sense close to our purported final theory.

4. At first sight it might appear that this endpoint is similar to that of biology prior to the discovery of evolution by natural selection. However, it is rather different. It is about the discovery of a complete form for the laws and true constants of Nature. But even if we knew them we could not predict all the states that could emerge from them.

5. This does not mean that the entire Universe has to be as it is in every respect. Two universes with the same laws and constants of Nature, and even the same initial conditions, will display different outcomes to those laws and different detailed evolution because of symmetry breaking and quantum uncertainty.

6. Carter, 'Large Number Coincidences and the Anthropic Principle', in *Confrontation of Cosmological Theories with Observational Data*, ed. M.S. Longair, Reidel, Dordrecht, 1974.

7. A.R. Wallace, *Man's Place in the Universe*, Chapman & Hall, London, 1903, p. 267.

8. M. Born, *Physics in My Generation,* Pergamon, London, 1956, p. 77.

9. S. Schaefer, *Independent,* 4 June 2000, p. 6.

10. A. Guth, 'The Inflationary Universe', *Phys. Rev.* D 23, 347 (1981); A. Guth, *The Inflationary Universe,* Addison Wesley, Reading, 1997.

11. See J.D. Barrow, *The Origin of the Universe,* Orion, London, 1994 for an account of these developments.

12. This is because the Universe contains irregularities.

13. The acceleration is so fast that only a very brief period, from 10^{-35} to 10^{-33} seconds is required to do this.

14. For a longer account of this problem see J.D. Barrow and J. Silk, *The Left Hand of Creation,* Basic Books, New York, 1983 and 2nd edn Penguin Books, London, 1995.

15. G. Smoot and K. Davidson, *Wrinkles in Time,* Morrow, New York, 1994. J.C. Mather and J. Boslough, *The Very First Light,* Basic Books, New York, 1996.

16. J.D. Barrow and F.J. Tipler, *The Anthropic Cosmological Principle,* Oxford University Press, Oxford, 1986.

17. A. Linde, 'The Self-Reproducing Inflationary Universe', *Sci. American* no. 5, vol. 32 (1994).

18. Prepared for the author by Rob Crittenden.

19. I must confess that I have always been puzzled by this justification for the study of history. It seems that most of the major problems in the world, from Northern Ireland to the Middle East, have arisen because people know too much history.

20. N. Ferguson (ed.), *Virtual History,* Perseus Books, New York, 1997.

21. D. Mackay, see J.D. Barrow, *Impossibility,* Oxford University Press, London, 1998 for a fuller discussion.

22. K. Amis, *The Alteration,* Penguin, London, 1988, which imagines the consequences of the English reformation never having happened.

23. L. Deighton, *SS-GB,* Jonathan Cape, London, 1978, in which, by February 1941, the British have surrendered, Churchill has been executed, King George VI is imprisoned in the Tower of London and the SS are running Britain from Whitehall.

24. R. Harris, *Fatherland*, Hutchinson, London, 1992.

25. J.L. Borges, *Labyrinths*, New Directions, New York, 1964, p. 19.

26. M. Oakeshott, quoted in N. Ferguson (ed.), *Virtual History*, Perseus Books, New York, 1997, pp. 6–7.

27. Ferguson, ibid., p. 6.

28. S. Blackburn, *Being Good*, Oxford University Press, London, 2001, pp. 72–3.

29. Ferguson, (ed.), *Virtual History*, Perseus Books, New York, 1997, p. 86.

chapter ten

New Dimensions

1. H. Reichenbach, *The Philosophy of Space and Time*, Dover, New York, 1958, pp. 281–2.

2. J.W. McReynolds, 'George's Problem', *Scripta Mathematica* 15, 2 (June 1949).

3. I. Kant, 'Thoughts on the True Estimation of Living Forces', in J. Handyside (transl.), *Kant's Inaugural Dissertation and Early Writings on Space*, University of Chicago Press, Chicago, 1929.

4. The gravitational force between two point masses is proportional to r^{-2}, where r is their separation in space.

5. This is also true for electrical or magnetic forces.

6. To see this, consider a mass located at a point. Now surround it by a spherical surface. The lines of force attracting towards the mass point in all dimensions intersect every point on the spherical surface. It is the area of this surface that tells us what the inverse-power of distance is that the force obeys. In three-dimensional space the spherical surface is two-dimensional and has an area proportional to the square of its radius. Likewise, in N-dimensional space the sphere has a surface area threaded by the force lines that is proportional to its radius to the (N-1)st power.

7. Portrait of Immanuel Kant, © AKG London.

8. I. Kant, quoted in C. Pickover, *Surfing through Hyperspace*, Oxford University Press, New York, 1999, p. 9.

9. Perhaps they should have made these discoveries much earlier. Imagine viewing triangles, lines and geometrical relations on a flat surface using a curved mirror. The Euclidean geometry will be distorted into that of a curved surface. But there will still be a one-to-one correspondence between the rules governing flat geometry and those in the distorted space, guaranteed by the laws of light reflection.

10. Criticising Mach's study of n-dimensional geometries.

11. The challenge of imagining life in two dimensions came before the challenge of thinking about four dimensions. Gauss imagined two-dimensional creatures, that he called 'bookworms', that lived on infinite sheets of paper. Helmholtz (1881) put the bookworms on the surface of a ball, thus giving them a world that was finite in extent but without any boundaries.

12. This idea has been reworked by several authors periodically ever since, adding more geometrical and topological sophistication each time; for example, *Sphereland* (1964) by Dionys Burger, *Planiverse* by Dewdney (Pan, London, 1984) and *Flatterland* (2001) by Ian Stewart.

13. Notably Johann Zollner and members of the Psychical Society, who were lampooned in Oscar Wilde's *The Canterville Ghost*.

14. J.C.F. Zollner, 'On Space of Four Dimensions', *Quarterly Journal of Science* (new series) 8, 227 (1878).

15. B. Stewart and P. Tait, *The Unseen Universe*, Macmillan, London, 1884 also. He was the founder of knot theory and recognised that 3-dim knots could be unknotted in a 4th dimension.

16. For an interesting essay on the relation between Conan Doyle and Holmes, see Martin Gardner, 'The Irrelevance of Conan Doyle', *Beyond Baker Street*, ed. M. Harrison, Bobbs-Merrill, New York, 1976; reprinted in M. Gardner, *Science: Good, Bad and Bogus*, Prometheus Books, New York, 1981, Chapter 9.

17. C. Hinton, *A Picture of Our Universe* (1884), see *Speculations on the Fourth Dimension: Selected Writings of Charles Hinton*, ed. R. Rucker, Dover, New York, 1980, p. 41.

18. James Hinton even had unusual medical views. He wrote a book entitled *The Mystery of Pain* in which he put forward the theory that 'all that which we feel as painful is really *giving* – something that our fellows are better for, even though we cannot trace it.' His son Charles later tried to create a mathematical formulation of this idea using higher-dimensional geometry and infinite series!

19. C. Hinton, *Dublin University Magazine* 1880. It was reprinted as a pamphlet with the title 'What is the Fourth Dimension: Ghosts Explained' by Swann Sonnenschein & Co. in 1884. Mr Sonnenschein was a devotee of Hinton's ideas and published nine more of his pamphlets in the next two years. They were then gathered together and published as a two-volume collection entitled *Scientific Romances*. Those that feature extra dimensions are reprinted in C. Hinton, *Speculations on the Fourth Dimension: Selected Writings of Charles Hinton*, ed. R. Rucker, Dover, New York, 1980.

20. C. Hinton, 'A Mechanical Pitcher', *Harper's Weekly*, 20 March 1897, pp. 301–2.

21. A.L. Miller, *Einstein, Picasso: Space, Time and the Beauty that Causes Havoc*, Basic Books, New York, 2001.

22. Pablo Picasso, *Portrait of Dora Maar*, 1937, © Succession Picasso/DACS 2002.

23. U. Eco, *Foucault's Pendulum*, Secker & Warburg, London, 1989, p. 3.

24. A. Einstein, *Ann. de Physik* 35, 687 (1911).

25. A. Einstein, Letter to Ilse Rosenthal-Schneider, 13 October 1945, translation and German original in I. Rosenthal-Schneider, *Reality and Scientific Truth: Discussions with Einstein, von Laue, and Planck*, Wayne State University Press, Detroit, 1980, p. 37.

26. G.E. Uhlenbeck, *American Journal of Physics* 24, 431 (1956). Uhlenbeck was a student of Ehrenfest.

27. His letter read as follows: 'My dear friends: Bohr, Einstein, Franck,

Herglotz, Joffé, Kohnstamm, and Tolman! I absolutely do not know any more how to carry further during the next few months the burden of my life which has become unbearable . . . Perhaps it may happen that I can use up the rest of my strength in Russia . . . If, however, it will not become clear rather soon that I can do that, then it is as good as certain that I shall kill myself. And if that will happen some time then I should like to know that I have written, calmly and without rush, to you whose friendship has played such a great role in my life . . . In recent years it has become ever more difficult for me to follow the developments in physics with understanding. After trying, ever more enervated and torn, I have finally given up in desperation. This made me completely weary of life . . . I did feel condemned to live on mainly because of the economic cares for the children. I tried other things but that helps only briefly. Therefore I concentrate more and more on the precise details of suicide. I have no other practical possibility than suicide, and that after having first killed Wassik. Forgive me . . . may you and those dear to you stay well.'

28. P. Ehrenfest, 'In what way does it become manifest in the fundamental laws of physics that space has three dimensions?', *Proc. Amsterdam Academy* 20, 200 (1917) and *Annalen der Physik* 61, 440 (1920).

29. Original watercolour by Maryke Kammerlingh-Onnes, courtesy AIP Emilio Segrè Visual Archives.

30. Mathematicians are used to this peculiarity as well. It is frequently the case that a general mathematical conjecture will be decided one way or the other in all dimensions of space except three. Here it is usually specially difficult to decide.

31. K. Kuh, *The Artist's Voice*, Harper & Row, New York, 1962, p. 42.

32. G.J. Whitrow, *Brit. J. Phil. Sci.*, 6, 13 (1955).

33. G.J. Whitrow, *The Structure and Evolution of the Universe*, Hutchinson, London, 1959.

34. This is a rather natural question to ask because if the velocity of

light is a fundamental constant of Nature, the same for all observers no matter where they are and how they are moving, then it means that there is a deep and fundamental link between space and time. Einstein's theories of gravity and motion have shown us the consequences of this link. As a result physicists now speak of 4-dimensional *space-time* rather than of space *and* time. This synthesis was first introduced by Hermann Minkowski in a lecture entitled 'Space and Time' to scientists in Cologne on 21 September 1908. He began with the announcement, 'Gentlemen! The views of space and time which I wish to lay before you have sprung from the soil of experimental physics, and therein lies their strength. They are radical. Henceforth space by itself, and time by itself, are doomed to fade away into mere shadows, and only a kind of union of the two will preserve independence.' They imagine space-time like a 4-dimensional block that can be sliced up in many possible ways, each of them equivalent to a different way of defining 'time'. This image of block space-time is an old one since it arises rather naturally from a God's-eye view of the world. Back in the thirteenth century, Thomas Aquinas wrote that 'We may fancy that God knows the flight of time in His eternity, in the way that a person standing on top of a watchtower embraces in a single glance a whole caravan of passing travellers.' T. Aquinas, *Compendium Theologiae*, quoted in P. Nahin, *Time Machines*, AIP Press, New York, 1993, p. 103. The term 'block universe' was introduced by the Oxford philosopher Francis Bradley in his book *Principles of Logic* (1883), written many years before Minkowski's introduction of the mathematical description of space-time and the Wellsian fantasy of time travel. He writes: 'We seem to think that we sit in a boat, and are carried down the stream of time, and that on the bank there is a row of houses with numbers on the doors. And we get out of the boat, and knock at the door of number 19, and, re-entering the boat, then suddenly find ourselves opposite 20, and, having then done the same, we go on to 21. And, all this while, the firm fixed row of the past and future stretches in a block behind us, and before us.' Einstein also seems to have held this view, in which the future is laid out ready-made before us and any differences between the past, the present and the future are mere illusions. Writing to the family of his oldest and closest friend, Michele Besso, a few weeks after Besso's death in 1955, Einstein pointed to the illusory nature of the past and the future, knowing that there would be no recovery from his own illness: 'And now he has preceded me briefly in bidding farewell to this strange world. This signifies nothing. For us believing physicists, the distinction between past, present, and future is only an illusion, even if a stubborn one.' See B. Hoffman, *Albert Einstein: Creator and Rebel*, New American Library, New York, 1972, p. 257.

35.   See J.D. Barrow and F.J. Tipler, *The Anthropic Cosmological Principle*, Oxford University Press, Oxford, 1986, chap. 4 and F. Tangerlini, 'Atoms in Higher Dimensions', *Nuovo Cimento* 27, 636 (1963); J.D. Barrow, 'Dimensionality', *Phil. Trans Roy. Soc.* A, 310, 337 (1983); I. Freeman, *American Journal of Physics* 37, 1222 (1969). L. Gurevich and V. Mostepanenko, *Phys. Lett.* A 35, 201 (1971); I. Rozental, *Soviet Physics Usp.* 23, 296 (1981).

36.   From J.D. Barrow, *The Book of Nothing*, Jonathan Cape, London, 2000, based on a diagram constructed by M. Tegmark, *Annals of Physics* (NY), 270, 1 (1998).

37.   An interesting general point about two-dimensional machines has been made by John S. Harris (Brigham Young University). He pointed out the remarkable similarity between planiversal mechanisms and steriversal gun design. Of the German Mauser military pistol he writes that, 'This remarkable automatic pistol has no pivots or screws in its functional parts. Its entire operation was through sliding cam surfaces and two-dimensional sockets. Indeed, the lockwork of a great many firearms, particularly nineteenth century arms, follows essentially planiversal principles.'

Quoted in A. Dewdney (ed.), *A Symposium on Two-dimensional Science and Technology*, unpublished, 1981, p. 181.

38. One sees a manifestation of this in mathematics, where dynamical systems only start to display complex and chaotic behaviour when their trajectories move in three dimensions. Only then can they wind around each other in complicated forms without intersecting.

39. J. Dorling, 'The Dimensionality of Time', *Am. J. Phys.*, 38, 539 (1969). F.J. Yndurain, 'Disappearance of matter due to causality and probability violations in theories with extra timelike dimensions', *Physics Letters* B 256, 15 (1991).

40. Ratibor is now in Poland, and renamed Raciborz.

41. T. Kaluza, 'Zum Unitätsproblem der Physik', *Sitzungsberichte Preussische Akademie der Wissenschaften* 96, 69 (1921).

42. O. Klein, *Zeit. f. Physik* 37, 895 (1926) reprinted and translated in O. Klein, *The Oskar Klein Memorial Lectures*, ed. G. Ekspong, World Scientific, Singapore, 1991, p. 103.

43. P. Candelas and S. Weinberg, *Nucl. Phys.* B. 237, 397 (1984).

44. E. Wharton, *Vesalius in Zante*, quoted in C. Pickover, *Surfing through Hyperspace*, Oxford University Press, New York, 1999, p. 118.

45. If there is more than one extra dimension then R is the mean size of all the extra dimensions.

46. For some simple descriptions of why this problem arises and why it is cured in string theories see J.D. Barrow, *Theories of Everything*, Oxford University Press, Oxford, 1991, pp. 22–3, 80–5 and M. Green, 'Superstrings', *Scientific American*, September issue (1986), p. 48.

chapter eleven

Variations on a Constant Theme

1. G.A. Cowan, *Scientific American*, vol. 235, July 1976 issue, p. 41.

2. P. Levi, *The Periodic Table*, Abacus, London, 1986 pp. 196–7. To

set the reader's mind at rest, the metal that Levi's colleague had in his possession turned out to be cadmium.

3. R. Bodu, H. Bouzigues, N. Morin and J.P. Pfifelman, 'Sur l'existence d'anomalies isotopiques rencontrées dans l'uranium d'Oklo', *Comptes Rendus Acad. Sci. Paris*, Series D 275, 1731 (1972).

4. This analysis was performed by mass spectrometry. Molecules of uranium hexafluoride gas are ionised and accelerated before being deflected when they pass through a magnetic field. The mass of the molecule is revealed by the magnitude of the deflection. The accuracy of this technique is very high. It was standard for the 'normal' natural abundance of uranium to be proscribed by 0.7202 $\pm$ 0.0006 per cent of uranium-235, whereas the samples analysed from Oklo showed 0.7171 $\pm$ 0.0007 per cent.

5. Isotopes are forms of the same element for which the nucleus contains the same number of protons but a different number of neutrons. The simplest example is that of hydrogen whose nucleus contains only one proton and no neutrons. Deuterium, the smallest isotope of hydrogen, contains one neutron and one proton.

6. The standard value is 0.007202 $\pm$ 0.00006.

7. Seven isotopes of neodymium are found. One of these, neodymium-142, is not a fission product and can be used to determine the abundances of the natural neodymium components at the Oklo site before they were affected by the running of the reactor.

8. M. Neuilly, J. Bussac, C. Frejacques, G. Nief, G. Vendryes and J. Yvon, 'Sur l'existence dans un passé reculé d'une réaction en chaîne naturelle de fissions, dans le gisement d'uranium d'Oklo (Gabon)', *Comptes Rendus Acad. Sci. Paris*, Series D 275, 1847 (1972).

9. P.K. Kuroda, 'On the Nuclear Stability of Uranium Minerals', *J. Chem. Phys.* 25, 81–2 (1956), and 'On the Infinite Multiplication Constant and Age of U Minerals', *J. Chem. Phys.* 25, 1295–6 (1956).

10. George Cowan reports that a less detailed prediction was made in 1953 by George Wetherill of UCLA and Mark Inghram of

University of Chicago. They studied a deposit of pitchblende (a thorium-poor form of the uranium oxide, uranimite that crystallises into a colloidal solution) and wrote that: '[Our] calculation shows that 10 per cent of the neutrons produced are absorbed to produce fission. Thus the deposit is 25 percent of the way to becoming a [nuclear reactor] pile. It is also interesting to extrapolate back 2000 million years, when the uranium-235 abundance was 3 per cent instead of .7. Certainly such a deposit would be closer to being an operating pile.' Quoted in *Sci. American*, vol. 235, July 1976 issue, pp. 40–1. The original article is G.W. Wetherill and M.G. Inghram, *Proc. Conf. Nucl. Processes Geol. Settings*, pp. 30–2, Nat. Research Council, Washington DC (1953).

11. The time of 1.84 ± 0.07 billion years ago for the onset of criticality (obtained using uranium-lead dating) is pinned down by the requirement that it be far enough in the past for the uranium-235 abundance to be large enough but not so early that no liquid water was present to produce the concentrated uranium oxide rich solution. The operating life span of the reactor was $2.29 \pm 0.7 \times 10^5$ years, see Y.V. Petrov, 'The Oklo Natural Nuclear Reactor', *Sov. Phys. Usp.* 20, 937 (1977) and J. M. Irvine, R. Naudet, 'The Oklo Nuclear Reactors: 1800 Million Years Ago', *Interdisciplinary Science Reviews* 1, 72 (1976).

12. A detailed reconstruction of events showed that about 1.8 billion years ago the peculiar geological layout of this part of Gabon facilitated the creation of self-sustaining chain reactions in six natural nuclear reactors. The total average power output over the 200,000 years of reactor activity is rather feeble, roughly 25 kilowatts.

13. M. Maurette, 'Fossil Nuclear Reactors', *Ann. Rev. Nucl. Sci.*, 26, 319 (1976); J.C. Ruffenach, R. Hagemann and E. Roth, 'Isotopic Abundance Measurements a Key to Understanding the Oklo Phenomenon', *Zeit Naturforsch.* 35A, 171 (1979).

14. H.G. Wells, *Tono-Bungay*, Waterlow & Sons, London, 1933, p. 215. This remarkable novel, first published in 1909, tells of a secret

adventure of the scientist explorer Gordon-Nasmyth to bring back radioactive material from West Africa – one of the daring enterprises of the tycoon Ponderevo whose miraculous remedy Tono-Bungay gives the book its title. Tons of 'festering' earth are loaded into their ship, worth its weight in gold, but the irradiation of the ship's wooden fibres causes it to leak. Eventually it sinks and the bankrupt venture capitalists are rescued by a Union Castle ocean liner.

15. Photograph courtesy of Ilya Shlyakhter; for further information, see the website http://alexonline.info.

16. Y. Fujii *et al.*, 'The Nuclear Reaction at Oklo 2 Billion Years Ago', *Nucl. Phys.* B 573, 381 (2000).

17. A.I. Shlyakhter, *Nature* 260, 340 (1976); A.I. Shlyakhter, *Direct test of the time-independence of fundamental nuclear constants using the Oklo natural reactor*, ATOMKI Report A/1, Leningrad Nuclear Physics Institute, 1983.

18. T. Damour and F. Dyson, *Nucl. Phys.* B 480, 37 (1996).

19. Y. Fujii *et al.*, 'The Nuclear Reaction at Oklo 2 Billion Years Ago', *Nucl. Phys.* B 573, 381 (2000).

20. Damour and Dyson's analysis can be interpreted as giving the ranges -94 ± 26 meV and 46 ± 44 meV which they choose to run together to create a single range bounded by the endpoints (this now includes zero which the separate ranges did not) of -120 meV $< \Delta E_r < 90$ meV.

21. Fujii *et al.*, op. cit., consider neutron capture by an isotope of gadolinium. This is a promising approach based upon new samples but the problem of contamination is acute and significant correction of the analysis is necessary to allow for it. The most reasonable choices appear to favour the right-hand branch solution for samarium, consistent with zero shift in the resonance energy in three of the four samples analysed.

22. Kaluza-Klein theories with extra dimensions of space, that we looked at in the last chapter, predict that α and α_s will both be

proportional to R^{-2}, where R is the average diameter of any extra dimensions of space, if R changes with time.

23. E. Teller, *Conversations on the Dark Secrets of Physics*, Plenum, New York, 1991, p. 87.

24. D.H. Wilkinson, *Phil. Mag.* (series 8) 3, 582 (1958).

25. F. Dyson, *Phys. Rev. Lett.* 19, 1291 (1967).

26. A. Peres, *Phys. Rev. Lett.* 19, 1293 (1967); S.M. Chitre and Y. Pal, *Phys. Rev. Lett.* 20, 278 (1968); T. Gold, *Nature* 218, 731 (1968).

27. *Observer*, 27 January 2002, p. 30.

28. F. Hoyle, *Comet Halley*, Michael Joseph, London, 1985.

29. J. von Neumann, *Collected Works*, ed. A. H. Taub, Pergamon, New York, 1961, vol. 6, article 39.

chapter twelve

Reach for the Sky

1. O. Wilde, *The Critic as Artist* (1890) in *The Portable Oscar Wilde*, eds R. Aldington and S. Weintraub, Viking, New York, 1976.

2. R. Browning, *The Poems of Robert Browning*, Heritage Press, New York, 1971.

3. G. Gamow, *Phys. Rev. Lett.* 19, 759 (1967). A measurement had been attempted by M.P. Savedoff, 'Physical Constants in Extra-Galactic Nebulae', *Nature* 178, 688–9 (1956).

4. J. Bahcall, W. Sargent and M. Schmidt, *Astrophys. J.* 149, L11 (1967).

5. R. Alpher, 'Large Numbers, Cosmology, and Gamow', *American Scientist* 61, 56 (1973). Reproduced by permission of *American Scientist*.

6. J. Bahcall and M. Schmidt, 'Does the Fine-Structure Constant Vary with Cosmic Time?', *Phys. Rev. Lett.* 19, 1294–5 (1967).

7. M.J. Drinkwater, J.K. Webb, J.D. Barrow and V.V. Flambaum, *Mon. Not. Roy. Astron. Soc.* 295, 457 (1998).

8. Actually it measures the constancy of the product $g_p \alpha^2$ where g_p

is the proton 'g factor'. We assume here that g_p is not changing.

9. Again we assume that g_p is constant.

10. L.L. Cowie and A. Songalia, *Astrophys. J.* 453, 596 (1995).

11. This limit excludes inclusion of uncertainties associated with possible variations of local velocities of the line sources.

12. These simulations have been developed to predict the locations of the spectral lines and energy levels of the atoms in the lab already and are carried out by Victor Flambaum and his colleagues at the University of New South Wales.

13. This improved sensitivity arises because the sensitivity to α with respect to relativistic aspects of atomic structure enters as $(\alpha Z)^2$ where Z is the atomic number (number of protons in the nucleus) of the atom. Thus by comparing lines of different atomic species with large and small values of Z a significant gain in sensitivity is obtained over methods which observe doublets of a species with the same Z.

14. W. Maudlin, cartoon caption *Up Front* (1945).

15. *Scientific American*, November 1998, Science and the Citizen 'Inconstant constants', quoting Robert J. Scherrer.

16. The measurement of the required spectral lines in the laboratory at the required level of accuracy (for which there appears to have been no need before) is very challenging and with more laboratory observations the MM method could extract even more information from the available data.

17. J.K. Webb, M.T. Murphy, V.V. Flambaum, V.A. Dzuba, J.D. Barrow, C.W. Churchill, J.X. Prochaska and A.M. Wolfe, 'Further evidence for cosmological evolution of the fine structure constant', *Phys. Rev. Lett.* 87, 091301 (2001). When new data are included from W. Sargent the statistical significance of the detection of variation in α is better than 7-sigma.

18. Prepared for the author by Michael Murphy.

19. This can be compared with the results obtained with the first round of observations in 1999:

$$\Delta\alpha/\alpha = [\alpha(z) - \alpha(now)]/\alpha(now) = (-1.09 \pm 0.36) \times 10^{-5}$$
published by J.K. Webb, V.V. Flambaum, C.W. Churchill, M.J. Drinkwater and J.D. Barrow, *Phys. Rev. Lett.* 82, 884 (1999).

20. A.S. Eddington, *New Pathways in Science*, Cambridge University Press, Cambridge, 1935, p. 211.

21. There are other forms of error which are introduced deliberately, especially by politicians, when treating voting data. For example, a party with a ten-point manifesto assumes without question that if they win the election by an overall majority then they have a mandate for all their manifesto policies whereas they might in reality only have a majority vote for a fraction of them.

22. This is the effect of refraction of the incoming light which depends upon the depth of atmosphere it has to traverse which depends on the geographical latitude of the telescope. It is a very small effect, usually ignorable in astronomy, but it enters at the same level as the apparent fine structure variations. If corrected for, it makes the value of the fine structure constant slightly smaller still in the past when compared with its value today.

23. J.D. Prestage, R.L. Tjoelker and L. Maleki, *Phys. Rev. Lett.* 74, 18 (1995).

24. In the future new atomic interferometers may offer an improvement on the Prestage limit. The current experimental resolution of this technology is sensitive to shifts in α of about 10^{-8} over 1–2 hours. In the future it may be adapted to test the constancy of α. There is no immediate prospect of it approaching astronomical levels of accuracy though. Motivated by new atomic physics calculations by V. Dzuba and V. Flambaum, *Phys. Rev.* A 61, 1 (2000), Torgerson has discussed the potential of optical cavities to provide improved measurements of α stability with time (see *Physics*/0012054 (2000)). He expects laboratory experiments soon to be sensitive to time variations of order 10^{-15} per year.

25. P.P. Avelino *et al.*, *Phys. Rev.* D 62, 123508 (2000) and R. Battye, R. Crittenden and J. Weller, *Phys. Rev.* D 63, 043505 (2001).

26. Since the sensitivity of the microwave temperature anisotropy observations is about 2×10^{-5} and last scattering occurs about 14 billion years ago, using our best age estimates, we could not obtain a limit on the time variation of α from this data that was better than $(2 \times 10^{-5})/(14 \times 10^{9} \text{ yrs}) \approx 1.4 \times 10^{-15}$ per year.

27. H. Mankell, *Sidetracked*, Harvill Press, London, 2000, p. 3.

28. The theory including varying G is the Brans-Dicke theory of gravity, found by Carl Brans and Robert Dicke, *Physical Review* 124, 924 (1961). A cosmological theory including varying α was found by Håvard Sandvik, João Magueijo and myself in 2001 (*Phys. Rev. Lett.* 88, 031302 (2002)), extending developments by Jacob Bekenstein, *Physical Review* 25, 1527 (1982).

29. It increases in proportion to the logarithm of the age of the Universe; for full details see J.D. Barrow, H. Sandvik and J. Magueijo, 'The Behaviour of Varying-alpha Cosmologies', *Physical Review* D 65, 063504 (2002).

30. J.D. Barrow, H. Sandvik and J. Magueijo, 'Anthropic Reasons for Non-zero Flatness and Lambda', *Physical Review* D 65, 123501 (2002).

chapter thirteen

Other Worlds and Big Questions

1. W. Owen, 'O world of many worlds', *The Collected Poems of Wilfred Owen, 1893–1918*, Chatto & Windus, London, 1963.

2. C. Pantin, 'Life and the Conditions of Existence', in *Biology and Personality*, ed. I.T. Ramsey, Blackwell, Oxford, 1965, p. 94; see also C.F.A. Pantin, 'Organic Design', *Advances in Science* 8, 138 (1951).

3. The problem of adding substance to the word 'probable' is a deep and difficult one. Every attempt to define probability for cosmological problems precisely and so give numerical answers to questions like 'what is the probability that the universe has certain

properties which will allow life to exist in it?' has so far failed. Technically, this is the mathematical problem of defining a probability measure. The difficulty is simply not knowing what are the equally likely outcomes when it comes to assessing the collection of all possible starting conditions for the universe or all possible outcomes from the chaotic inflationary universe theory. The problems are exacerbated by the problem of defining 'when' the probabilities apply in a way that is universal for every place in the universe. There is considerable research into this problem at present but it remains unsolved.

4. C. Pantin, 'Life and the Conditions of Existence', op. cit. p. 104. Note that although Pantin mentions 'a solution analogous to the principle of natural selection' he does not develop it.

5. We could be wrong about this if the Theory of Everything contained some inter-linkage of constants that had the property that a change of a part in a hundred billion in the fine structure constant produced a change of, say, one part in two of some other life-critical constant.

6. If life is nothing more than a by-product of a very high level of complexity being attained, then maybe there can be life in velocity space or in the fabric of space-time structure or down on the atomic, nuclear or elementary particles scales as an asymptotic extension of present quests to create nanotechnologies.

7. J.D. Barrow, *Pi in the Sky*, Oxford University Press, Oxford and Vintage, New York, 1992, pp. 280–92. For a development see also M. Tegmark, 'Is the "Theory of Everything" merely the Ultimate Ensemble Theory?', *Annals of Physics* (NY) 270, 1 (1998).

8. We could ask if there is some threshold in complexity at which it becomes possible for life to be described within a mathematical formalism. The only noticeable threshold occurs when we reach the complexity of arithmetic. At this point self-reference is possible. There can be a one-to-one correspondence between arithmetic and statements about arithmetic (this is not possible with simpler

structures like geometries). Cellular automata like John Conway's game of life turn out to be equivalent to arithmetic in their logical structure. It is interesting that when we attain the complexity of arithmetic the property of Gödel incompleteness becomes a property of the system. Some writers, notably John Lucas and Roger Penrose, have suggested that this property might be an essential feature of consciousness. If so, then the complexity threshold that is passed when we reach arithmetic would be the minimum level needed for conscious information processing to arise within the logical system. It is interesting to compare this low threshold for self-referential complexity in logical systems with the low threshold for generating complexity in discrete cellular automata discussed by Stephen Wolfram in *A New Kind of Science*, Wolfram Media Inc., Champaign, IL, 2002. Simple one-dimensional algorithms with nearest-neighbour rules can generate levels of complexity that are unsurpassed by adding extra dimensions, more complex rules, random perturbations or averaging.

9. If any false statement is present in a logical system it can be used to prove the truth of any statement (like $0 = 1$). Famously, Bertrand Russell responded to a challenge to show that his questioner was the Pope if $1 = 2$: you and the Pope are two but if two equals one then you and the Pope are one.

10. G.H. Hardy, *A Mathematician's Apology*, Cambridge University Press, Cambridge, 1967, p. 135.

11. This faith that the future will be like the present is what philosophers call the problem of induction.

12. W. Allen, *Getting Even*, Random House, New York, 1971, p. 33.

13. If we forget about inflation as a creator of diversity and just suppose that the Universe is infinite and random then somewhere, infinitely often, there must arise large regions which have life-supporting properties. We would have to inhabit one of them. However, the large ordered regions would be far less probable than small ones and inflation offers a mechanism to explain why

large ordered regions get generated with high probability.

14. A. C. Clarke, 'The Wall of Darkness', in *Super Science Stories*, collected in *The Other Side of the Sky*, Signet, New York, 1959, ch. 4. This story was written in 1946 and first published in 1949.

15. A. Linde, 'The Self-reproducing Inflationary Universe', *Scientific American* 5, 32 (May 1994).

16. The motivation for discovering if this is possible is to avoid doing it by accident.

17. E. R. Harrison, 'The Natural Selection of Universes Containing Intelligent Life', *Quarterly Journal of the Royal Astronomical Society* 36, 193 (1995). Although the author calls the intelligent tuning of the constants of Nature 'natural selection' of universes it is actually 'unnatural' selection or the 'forced breeding' of universes with desired characteristics.

18. L. Smolin, *The Life of the Cosmos*, Oxford University Press, New York, 1995.

19. This idea that the constants of Nature are 'reprocessed' when matter collapses into a singularity of infinite density, for example when a closed universe collapses and bounces back into a state of expansion, was first suggested by John A. Wheeler; see for example the last chapter of C. Misner, K. Thorne and J.A. Wheeler, *Gravitation*, W.H. Freeman, San Francisco, 1972.

20. This is rather like the rat-race long-term state for an evolutionary system, whereas the situation where a local maximum is achieved with respect to the value of the constants is like the attainment of an evolutionarily stable strategy, in which any deviation from this state leaves at least one of the players worse off; see for example J. Maynard Smith, *Evolutionary Genetics*, Oxford University Press, London, 1989.

21. Clearly, this scenario requires the universe to be closed so that it can collapse in the future.

22. Assuming that there is no other way for the constants to change their values other than by changing at a singularity.

23. The total energy of the Universe in any cycle is actually zero.

24. This was first noticed in two articles by the American cosmologist R.C. Tolman, 'On the Problem of the Entropy of the Universe as a Whole', *Physical Review* 37, 1639 (1931) and 'On the Theoretical Requirements for a Periodic Behaviour of the Universe', *Physical Review* 38, 1758 (1931).

25. J.D. Barrow and M. Dabrowski, 'Oscillating Universes', *Monthly Notices of the Royal Astronomical Society*, 275, 850 (1995).

26. C. Raymo, *Skeptics and True Believers*, Random House, New York, 1999, p. 221.

Index